AF716365

# Mining, Materials, and the Sustainable Development Goals (SDGs)

# Mining, Materials, and the Sustainable Development Goals (SDGs)

## 2030 and Beyond

Edited by
Cristian Parra, Brandon Lewis, and Saleem H. Ali

CRC Press
Taylor & Francis Group
Boca Raton London New York

CRC Press is an imprint of the
Taylor & Francis Group, an **informa** business

First edition published 2021
by CRC Press
6000 Broken Sound Parkway NW, Suite 300, Boca Raton, FL 33487-2742

and by CRC Press
2 Park Square, Milton Park, Abingdon, Oxon, OX14 4RN

© 2021 Taylor & Francis Group, LLC

CRC Press is an imprint of Taylor & Francis Group, LLC

Reasonable efforts have been made to publish reliable data and information, but the author and publisher cannot assume responsibility for the validity of all materials or the consequences of their use. The authors and publishers have attempted to trace the copyright holders of all material reproduced in this publication and apologize to copyright holders if permission to publish in this form has not been obtained. If any copyright material has not been acknowledged please write and let us know so we may rectify in any future reprint.

Except as permitted under U.S. Copyright Law, no part of this book may be reprinted, reproduced, transmitted, or utilized in any form by any electronic, mechanical, or other means, now known or hereafter invented, including photocopying, microfilming, and recording, or in any information storage or retrieval system, without written permission from the publishers.

For permission to photocopy or use material electronically from this work, access www.copyright.com or contact the Copyright Clearance Center, Inc. (CCC), 222 Rosewood Drive, Danvers, MA 01923, 978-750-8400. For works that are not available on CCC please contact mpkbookspermissions@tandf.co.uk

*Trademark notice*: Product or corporate names may be trademarks or registered trademarks, and are used only for identification and explanation without intent to infringe.

ISBN: [978-0-367-35850-1] (hbk)
ISBN: [978-0-367-81496-0] (ebk)

Typeset in Times
by Deanta Global Publishing Services, Chennai, India

# Contents

# Preface

In September 2015, the 193 member states of the UN General Assembly collectively adopted a comprehensive global agenda for the next 15 years: the Sustainable Development Goals (SDGs), also known as Agenda 2030. While the goals were adopted by governments, they represent the aspirations of humankind. Similarly, they will not be achieved without the collective efforts of humankind, including and especially requiring the efforts and financial wherewithal of the private sector. Mining is the ultimate primary industry for it forms the basis of most infrastructure development and even provides minerals for fertilizers and hence food production. Yet, the impact and benefits of mining are often hidden from the end consumer, making it difficult for the general public to evaluate its primacy in development.

This book is being published around five years after the launch of the SDGs and will thus be able to provide a mid-range evaluation of what has been accomplished and what needs to be done to ensure a better understanding of the linkages between mining and the targets under each goal. We are interested in highlighting both the enabling and constraining factors that mining brings forth to the SDG agenda. Extractive industries tend to polarize scientists as well as policy-makers and a dispassionate analysis is often lacking in the general discourse around the sector. Our aim in this anthology is to provide an evidence-based analysis of mining's linkages to the SDG agenda.

The book has been deliberately crafted chapter by chapter for each of the 17 goals to make it easier for decision-makers in industry and government to use while also providing analytical clarity for academic observers and students. In the tradition of CRC Press books, we are providing a hands-on applied narrative in each of the chapters rather than theoretical deliberations. That is not to undermine the role of theory but rather to highlight the audience we have in mind for this volume. Case examples are also provided where possible to further animate the narrative. We have tried to include wide geographic coverage in the examples presented as well as a diversity of author profiles from academia, industry, and civil society. Such a multi-stakeholder approach to developing this anthology also gives the text more immediate relevance to a broad audience.

Mineral resource governance is on the front burner of the United Nations Development Programme and related organizations. The publishing of the Mineral Resource Governance Assessment by the International Resource Panel in 2020, coupled with the advent of a major consultation process being underway to inform the United Nations Environment Assembly in 2021, provides timely perspective to this volume. The 2020 Coronavirus Pandemic has made the world realize how interconnected global supply chains are for all commodities. Mining is at the base of so many supply chains and no doubt this book's relevance will also acquire additional salience as we emerge from this most debilitating time for the global economy in almost a century.

**Cristian Parra**
**Brandon Lewis**
**Saleem H. Ali**

# Editors

**Cristian Parra** is a Chilean–Australian economist at Malthus Global with 20 years of professional experience working with the extractive industry and its stakeholders globally. He has a long track record of senior advising in areas related to socioeconomic issues and progress of mining regions, social and economic mining performance, and impact assessment in developing and developed countries.

Cristian has led and participated in projects in 15 countries in more than 50 mining regions across Latin America, Asia-Pacific, and Africa, working with important multinational resources companies, development institutions, donors, governments, NGOs, and community groups.

**Brandon Lewis** is a sustainable investment professional with over a decade of global experience working at the intersection of natural resources, finance, and policy. He has worked on four continents in fields as wide-ranging as mining, renewable energy, forestry, agriculture, project finance, microfinance, and policy. He holds a BSc in Geology, an MSc in Economic Geology, and an MPA in International Development. In 2015 at the World Economic Forum, in collaboration with UNDP and the Columbia Center on Sustainable Investment, he co-authored *Mapping Mining to the Sustainable Development Goals: An Atlas.*

**Saleem H. Ali** holds the Blue and Gold Distinguished Professorship in Energy and the Environment at the University of Delaware, Newark, Delaware where he also directs the Gemstones and Sustainable Development Knowledge Hub. He is also a Senior Fellow at Columbia University's Center on Sustainable Investment and retains professorial affiliation at the University of Queensland's Sustainable Minerals Institute in Brisbane, Australia. Professor Ali earned his doctorate in Environmental Planning from Massachusetts Institute of Technology (MIT), Cambridge, Massachusetts; a Master's degree in Environmental Studies from Yale University, New Haven, Connecticut; and a Bachelor's degree in Chemistry from Tufts University, Boston, Massachusetts (summa cum laude). His books include *Treasures of the Earth: Need, Greed and a Sustainable Future* (Yale University Press); *Environmental Diplomacy* (with Lawrence Susskind, Oxford University Press), and *Mining, the Environment and Indigenous Development Conflicts* (University of Arizona Press).

# Contributors

**Saleem H. Ali**
Delaware University
Newark, Delaware

**P.B. Anand**
Bradford Centre for International Development
University of Bradford
London, United Kingdom

**Paul Cordy**
Cordy Geoscience
Squamish, British Columbia

**Kirsten Dales**
Industrial Ecologist and Advisor
Migori, Kenya

**Graham A. Davis**
The Payne Institute for Public Policy
Colorado School of Mines
Golden, Colorado

**Jorden Steven de Haan**
Pact Mines to Market
Nairobi, Kenya

**Stephen D'Esposito**
Resolve
Washington, District of Columbia

**Janet Fishlock**
Senior Social Development Specialist
Toronto, Ontario

**Laura García Jaramillo**
Senior Advisor Extractive Industry
Colombia

**Damien Giurco**
Institute for Sustainable Futures
University of Technology Sydney
Sydney, Australia

**Gavin Hilson**
Surrey Business School
University of Surrey
Surrey, United Kingdom

**Andrea Juárez**
Consultoría y Tecnología Ambiental
Ciudad de Guatemala, Guatemala

**Susan Keane**
Planet Gold
Washington, District of Columbia

**Joseph Kirschke**
International Copper Association
Washington, District of Columbia

**Daniele La Porta**
The World Bank
Washington, District of Columbia

**Estelle Levin-Nally**
Levin Sources
Cambridge, United Kingdom

**Natalie Lowrey**
Deep Sea Mining Campaign
Melbourne, Australia

**Victoria R. Nalule**
Extractives Hub
CEPMLP
University of Dundee
Dundee, United Kingdom

**Karin E. Olson Hoal**
Cornell University
Ithaca, New York

**Jennifer Peyser**
Resolve
Washington, District of Columbia

**Cristian Parra**
Malthus Global
Brisbane, Australia

**Blanca Racionero Gómez**
Levin Sources
Girona, Catalonia, Spain

**Ben Radley**
Development Studies
London School of Economics
London, United Kingdom

**Rodrigo Rodriguez-Fernandez**
NCD Asia Pacific Alliance
Tokyo, Japan

**Helen Rosenbaum**
Deep Sea Mining Campaign
Melbourne, Australia

**Nicole M. Smith**
Colorado School of Mines
Golden, Colorado

**Samantha Smith**
Blue Global Solution
Toronto, Ontario

**John Steen**
Bradshaw Research Initiative in Minerals and Mining (BRIMM)
The University of British Columbia
Vancouver, British Columbia

**John F.H. Thompson**
PetraScience Consultants
Vancouver, British Columbia

**Francesca Viliani**
International SOS
Copenhagen, Denmark

**John Weakliam**
VITA
Dublin, Ireland

**Mirko S. Winkler**
Swiss Tropical and Public Health Institute
Basel, Switzerland

# Introduction

The purpose of this book is to cast a vision – that of a mining sector that contributes evermore significantly to achieving the United Nations Sustainable Development Goals (SDGs) and Agenda 2030. This introduction begins by defining critical terms, before taking a brief look at mining and sustainability since 2015. It then explores growing trends in the field, before finishing with an overview of the chapters that follow.

## WHOSE SUSTAINABILITY?

Before elaborating further, a brief comment on how we define sustainability. As the concept becomes more widespread, definitions proliferate. This can create confusion and disagreement about what sustainability means, and sometimes even desensitization to its importance. In other words, when sustainability means everything, it means nothing. In this book, despite the fact we recognize the growing complexity of the global sustainable development challenges, we define sustainable development in the original language of the 1987 Brundtland Report, as "meeting the needs of the present without compromising the ability of future generations to meet their own needs."* Sustainability, then, is the discipline of ensuring that the same environmental, social, biological, and economic endowments bestowing good on the earth and its inhabitants today remain healthy and able to bestow the same goods on the earth and its inhabitants in the future.

Two dichotomies may prove useful in understanding the different concepts in this book: those of weak versus strong sustainability, and manufactured versus natural capital. Natural capital is essentially the value and productive capacity of nature, otherwise known as ecosystem services. Manufactured or human capital is essentially the value and productive capacity of technology. Weak sustainability presumes full substitutability of human (or manufactured) capital for natural capital. In a weak sustainability paradigm, pollution is not problematic as long as it can be cleaned up, and ecosystems can be compromised as long as such negative impacts are "offset" through positive activities elsewhere. Strong sustainability separates the two types of capital, arguing that manufactured capital cannot substitute for natural capital. Strong sustainability paradigms point out that species extinction, for example, is irreversible, and that the complex interdependencies of natural systems often means there is a high level of uncertainty about the impacts of degrading natural capital.†

While this book does not take a normative stance on which paradigm to use, it is worth noting that many differences in perspective about sustainability and sustainable development can be traced back to these two dichotomies. It is also worth noting

* https://www.are.admin.ch/are/en/home/sustainable-development/international-cooperation/2030agenda/un-_-milestones-in-sustainable-development/1987--brundtland-report.html

† https://sustainabledevelopment.un.org/content/documents/6569122-Pelenc-Weak%20Sustainability%20versus%20Strong%20Sustainability.pdf

that over time, as the chapters that follow demonstrate, the world seems to slowly be shifting from a weak to a strong sustainability perspective.

## A FIVE-YEAR RETROSPECTIVE

Five years ago, following interviews with more than 60 global experts representing the mining industry, government, NGOs, academia, and community groups, about mining and sustainable development, the World Economic Forum, Columbia Center on Sustainable Investment, UN Development Programme, and UN Sustainable Development Solutions Network published the report "Mining and the SDGs: An Atlas." The "Atlas" explored the ways in which the mining industry could contribute to the SDGs, as well as the ways in which the SDGs would drive change within the mining industry.

That year, 2015, was the year Agenda 2030 (formalized in the SDGs) was adopted. It was the year of the Paris Accord. The United Nations and the global community writ large had learned lessons from the attempt to achieve the Millennium Development Goals over the previous 15 years. The SDGs represented a more democratic vision of "the world we want," informed by thousands of surveys and interactions with people all around the world from all walks of life. They were comprehensive. Energy was high, ambition was great, and enthusiasm was contagious. That same trend and sentiment was apparent in the mining industry, and it was reflected in the "Atlas."

Fast-forward five years, and it's now 2020. The Intergovernmental Panel on Climate Change (IPCC) no longer refers to climate change, but to the climate emergency. We're told we have 11 years left to drastically reduce greenhouse gas emissions before extreme negative impacts of climate change are locked in and become unavoidable.* Despite increasing deployment of renewables globally, we're told that emissions in 2019 were the highest ever recorded. We're losing biodiversity at the fastest rate ever recorded, and ecosystems are in grave danger of being lost for good. Inequality is growing and with it isolationism, xenophobia, and political unrest – including in previously relatively stable nations.

This is not a very inspiring picture. But where is the mining industry in all of this? What have the past five years looked like for mining, and what might the next ten look like? How has mining played a role in the triple threat of climate emergency, increasing inequality, and disappearing biodiversity? Has it been part of the problem, part of the solution, or both?

It's a mixed picture. Perhaps the two biggest stories relating to mining and sustainable development since 2015 are both negative ones involving two instances of the same issue with the same company – catastrophic and fatal tailings dam collapses at Vale's Samarco and Brumadinho mines in Brazil.† Hundreds of people were killed, entire communities were destroyed and displaced, the company's operations and output were halted for weeks, and physical damages mounted well into the billions of dollars.

---

* https://hbr.org/2018/12/the-story-of-sustainability-in-2018-we-have-about-12-years-left?referral=03759&cm_vc=rr_item_page.bottom

† www.ft.com/content/8c2f26f6-72b0-11e9-bf5c-6eeb837566c5

But that's not the whole story. Following the event at Brumadinho, partnerships were formed around a technical engineering issue the likes of which the world has never seen. Mine tailings dams shifted almost overnight from being an arcane engineering structure to a profoundly important social issue. The Church of England and Swedish AP Funds Council of Ethics rapidly took initiative, convening groups of NGOs, investors, and public figures form the Mining and Tailings Safety Initiative, requiring mining companies to disclose the safety and structural status of their tailings dams.* How might similar Partnerships for the Goals (to borrow language from SDG 17) form more proactively over the next ten years? What or who might drive their development, so that they form not in response to disasters but rather in response to collective opportunities?

## TRENDS IN MINING AND SUSTAINABILITY

### Sustainable Investing

Another powerful trend impacting the mining industry – in fact all industries – is the global rise in ESG (Environmental, Social, and Governance) Investing. This is the idea that incorporating analysis of a potential investment's environmental, social, and governance factors in addition to analysis of traditional financial metrics may help improve returns. The logic is that non-financial factors can have financially material impacts. Counter to the traditional idea in finance that the goal of business is to increase profits (à la Milton Friedman, 1962),† Larry Fink's 2018 letter to shareholders (CEO of Black Rock, the world's largest asset manager, overseeing $6 trillion in investments) rocked the investment world by arguing that the goal of business is to have a social purpose.‡

Is this a shot across the bow for mining companies, or an invitation to reframe or reposition their businesses in terms of achieving the collective societal good? To a large degree this will depend on how mining companies view themselves, whether as extractors of material from the earth answering only to shareholders, or as providers of material that will help the world respond to global challenges, answering to shareholders, employees, communities, and wildlife? Who are the stakeholders to whom the industry is ultimately responsive?

### Responsible Sourcing

Demands on the mining industry to fulfil social purpose are not only top-down, from the likes of asset managers and investors, but bottom-up, from the end users of mined materials. As societal awareness of the climate emergency, increasing inequality, and disappearing biodiversity grows, people are increasingly interested in where their materials come from, and technology is making it easier and easier for us to find this out. This is true of food, electronics, timber, clothing, and jewellery,

* www.churchofengland.org/more/media-centre/news/finance-news/thousands-tailings-dams-revealed-first-time-over-200-companies

† https://hbr.org/2012/04/you-might-disagree-with-milton

‡ www.blackrock.com/corporate/investor-relations/larry-fink-chairmans-letter

to name a few. Responsible sourcing commitments and traceability schemes are growing. Consumer expectations now include not only the quality of a product, but the nature of the processes used to create it. As these expectations grow and raw material provenance rises in importance, product differentiation becomes possible. Are the days of minerals as undifferentiated commodities over? Will sustainability certifications become necessary for market access, or will certified minerals command a higher market price? Will fair-trade minerals become as abundant and as well-understood of a concept as fair-trade coffee, organic produce, or sustainable forestry products?

Mined materials are essential to life on this planet, and have been since the dawn of humanity. Moreover, it is clear that mining vast quantities of particular minerals such as copper and rare earth elements (REEs) will be necessary to deploy renewable energy technologies at the scale required for substantial progress against greenhouse gas reduction goals. But the days of justifying a mining venture on the basis of a general justification for society's needs for a material are over. Providing raw materials is important, but the more relevant question in our times is not whether such materials can be provided, but rather at what cost – inclusive of environmental and social factors – they can be provided. The mined products are important, but so are the mining processes that get us those products. A company mining copper or REEs that end up in solar panels, wind turbines, other renewable electric generation technology, or transmission and distribution infrastructure can build on the positive societal impact by mining those products responsibly, or it can erase those positive impacts through deforestation, pollution, or operations that are unsafe for employees or worsen societal inequality.

## Unconventional Partnerships

One might argue that this is the *perfect storm* of factors, creating opportunities for innovative mining companies to differentiate themselves by using more sustainable production methods. As ore grades decline over time (the low-hanging fruit has already been discovered),* it is conceivable that major discoveries in mining will increasingly be not of new deposits, but of new technologies. Witness the partnership between Alcoa and Rio Tinto in 2018 to form Elysis, a joint-venture (partially backed by Apple)† to produce carbon-free aluminum; Apple purchased the first carbon-free commercial aluminum batch from Elysis on 5 December 2018.‡ The venture is significant in many ways:

- the improvement over a chemical engineering process (electrolysis of alumina) that had been the only option for aluminium smelting since it was invented in 1886
- the development of a process that emits oxygen instead of carbon dioxide

* https://hbr.org/2018/04/why-mining-yes-mining-cares-about-sustainability
† www.greentechmedia.com/articles/read/apple-new-joint-venture-announces-zero-carbon-aluminum-smelting
‡ www.elysis.com/en/apple-purchases-from-elysis-a-first-carbon-free-commercial-aluminium-batch

- the fact that the process reduces operating costs, increases production, and reduces greenhouse gas emissions simultaneously
- the partnership of two competitors (and two of the world's largest mining companies) in developing the process
- the mining companies' collaboration with and funding from a tech company

Where might future, similar developments occur within or alongside the mining industry in the future?

## Impact Measurement and Science-Based Targets

Corporate sustainability reports are replete with case studies and anecdotes illustrating genuinely positive contributions to sustainable development, and bottom-up key performance indicators are being developed to align with some SDG indicators. Significantly, there has been a proliferation of initiatives, standards, and reporting codes aimed at helping ensure mining contributes to the SDGs.

One study, "Leveraging great impact of mineral sustainability initiatives: An assessment of interoperability,"* mapped intersections between no fewer than 18 sustainability initiatives in mining: Aluminum Stewardship Initiative (ASI); Conflict-Free Gold Standard (CFGS); Conflict-Free Smelter Program (CFSP); Dodd-Frank Act (DFA); European Regulation on self-certification of responsible importers of tin, tantalum and tungsten, their ores, and gold originating in conflict-affected and high-risk areas (EUSC); Extractive Industries Transparency Initiative (EITI); Fairmined Standard for Gold and Associated Precious Metals (Fairmined); Fairtrade Standard for Gold and Associated Precious Metals (Fairtrade); Global Reporting Initiative – Mining and Metals Sector Supplement (GRI-M); Initiative for Responsible Mining Assurance (IRMA); International Council on Mining and Metals – Sustainable Development Framework (ICMM-S); International Cyanide Management Code (ICMC); International Standards Organization 14001 (ISO 14001); Mineral Certification Scheme of the International Conference on the Great Lakes Region (ICGLR); OECD – Due Diligence Guidance for Responsible Supply Chains of Minerals from Conflict-Affected and High-Risk Areas (OECD-D); Responsible Jewellery Council Code of Practices (RJC); Social Accountability SA 8000 (SA8000); and, of course, the UN SDGs.

However, currently no top-down framework exists to assess whether or not the cumulative effect of such contributions is significant enough to meaningfully help achieve the SDGs.† No standardized methods of impact measurement currently exists that can answer the following critical questions: How significant are mining companies' contributions to the SDGs to date? Are the contributions on pace to help the sector play a measurably meaningful role in the global accomplishment of Agenda 2030?‡

* www.csrm.uq.edu.au/media/docs/1520/giz-interoperability-report-july-2017-v3-17-aug-online.pdf

† Reference B Lewis and P Toledano GSS "Innovation"

‡ Reference B Lewis and P Toledano GSS "Innovation"

In other words, many companies can show site-level contributions to the SDGs, but what about national and global contributions?* Would it be possible for mining companies to determine their "responsibility" (based on revenue, profit, production volume, or other metrics) for contributing to clean water, emissions reductions, renewable energy generation, or gender equality? These are the types of questions that will become increasingly important over the next decade and beyond, and which this book makes an initial attempt to answer.

* https://smi.uq.edu.au/article/2018/04/can-we-measure-contribution-mining-sdgs

# GOAL 1 End Poverty in All Its Forms Everywhere

*Graham A. Davis*

## CONTENTS



## SUMMARY

SDG 1 seeks to eliminate the global incidence of extreme poverty, in all its forms and dimensions, by 2030. Income poverty is the aspect of poverty that is most widely measured, and which provides a definitive target against which to measure progress. Given that business-as-usual policies are expected to reduce income poverty from its current rate of 8.6% of the world's population to only 6% by 2030, achieving zero income poverty by then will require directed poverty-reducing strategies. Mining's most promising role within this strategy is through an increased demand for local inputs via backward linkages at existing and new mine sites.

## ANALYSIS OF MINING'S CONTRIBUTION TO INCOME POVERTY REDUCTION

The United Nations (UN) estimates that as of 2015, 736 million people were living in extreme income poverty. Extreme income poverty is defined as subsisting daily on less than what $US 1.90 would buy in goods and services in the United States at 2011 prices.

Extreme poverty rates have been falling over time. In 1990 36% of the world's population lived in extreme poverty. By 2018 this had been reduced to 8.6%. By 2030 the UN projects that 6% will live in extreme poverty. SDG 1 simply accelerates this natural rate of decline to zero. This is an ambitious task.

There are two main forces by which global poverty can be reduced. The first is economic growth. The second is a narrowing of the inequality of income distribution, such that income from economic activity is increasingly distributed away from the rich and towards the poor, usually via taxation and redistribution. The historical reductions in poverty have come mainly from past income growth, and not from increased redistribution. If anything, economic growth has brought increasing income inequality in developing countries over the past decade, and as a result, 61% of those in extreme poverty now live in middle-income countries like India and Nigeria.

Historically, a 1% increase in real GDP per capita has translated into about a 0.7% reduction in extreme poverty. This 0.7 conversion factor is known as the income elasticity of poverty. Historically, the world's developing economies have grown at an average of 1.7% per year, resulting in an annual reduction in the poverty rate of about 1.2% (note that this is percent change, not percentage points of the poverty rate). Progress has been slow. The remarkable drop in the global poverty rate since 1990 is mainly because of the rapid economic growth in China, which housed much of the world's poor. China also had an above-average income elasticity of poverty.

Most of the remaining extreme poverty exists in Southern Asia (SA) and Sub-Saharan Africa (SSA). This is no coincidence. Historically these two regions have had slow annual economic growth (1.9% and 0.2% respectively) and the lowest income elasticity of poverty of all regions (0.6 and 0.5 respectively). Their reduction in poverty over time has as a result lagged behind other regions.

The low growth in these two regions, and that growth's limited impact on poverty, implies a difficult path going forward. Based on historical performance, I estimate that their per capita real economic growth would have to accelerate to about 5% per year from 2019 to 2030 just to obtain the UN base case of 6% living in extreme poverty by 2030. Based on actual growth rates since 2015 SA may well achieve this, but SSA will not. A substantially higher rate of growth in both regions would be needed to reduce poverty to zero by 2030. This clearly will not be met. In other words, these regions cannot "grow" their way to the no-poverty goal of SDG 1. Direct provision of cash and services to the poor, funded both via domestic redistribution and foreign aid, is the only way to reach the goal of no poverty by 2030. This is a mandate that falls squarely on the domestic state, with a further role for the international community in building domestic capability and accountability in this regard.

The private sector, nevertheless, can play a supporting role. To examine mining's role in SDG 1, I note first that artisanal and industrial mining's contribution to historical growth-led reduction in poverty has been relatively muted. Most mining economies exhibit a temporary boom in income per capita that corresponds with the resource extraction cycle. During the boom poverty will reduce, while it will rise again during the subsequent bust. What matters for poverty reduction is long-run impacts. Estimates suggest that a typical mature mining economy like Chile or Peru has a level of per capita income about 25% higher than it would have been without mining, probably from productivity spillovers from the resource sector to the non-resource sectors. However, there is evidence that the income elasticity of poverty reduction is low in mining economies, resulting in no empirical evidence that poverty rates are lower than normal in mining economies despite their higher-than-normal levels of income per capita. That is, while mining has accelerated long-run economic growth, that accelerated growth has not led to measurable poverty reduction. It is for this reason that most empirical studies of mining and poverty find that poverty is higher than expected in economies with extensive mining activity. Based on this history, introducing mining into an economy, or expanding mining output where it is already present, will not lead to poverty-reducing economic growth.

There is very little mining in SA, due to both lack of geological endowments and an unfriendly investment climate towards Western investors. The prospect of increased mining investment in this region is slim, notwithstanding China's increased interest in the region. Mining's contribution to SSA's poverty reduction is more likely given the intensity of mining in the region. In 2009 the African Union crafted the Africa Mining Vision, with both short-term and long-term goals for mobilizing the benefits of mining, mainly through the development of resource-based development corridors. But as in SA, accelerated Western investment in SSA by 2030 is unlikely given the increasing difficulty of doing business in the region due to corruption, instability of fiscal regimes, and lack of security of property rights.

Recognizing the limits to investment and growth-led poverty reduction, the development community has targeted redistribution of resources to the poor, either via direct cash payments or the provision of health, education, and credit services, as the best path to accelerated poverty reduction. Within this framework a reasonable means by which mining can assist with poverty reduction is via fiscal returns to national governments. The International Council on Mining and Metals (ICMM), a mining industry organization "dedicated to a safe, fair and sustainable mining and metals industry," estimates that the largest mining companies in the world collectively pay about $25 billion in taxes and royalties to 50 national governments every year. These governments likely already use some of this funding to reduce poverty. An accelerated rate of poverty reduction would require an accelerated rate of transfers to the poor. If each of the 732 million people living in extreme poverty were to obtain an additional transfer of $1/day, that would require annual transfers to the poor of at least $255 billion (assuming direct cash transfers to the poor with no efficiency losses and low transactions costs). Not all of the $25 billion in mining tax receipts is generated in SA and SSA, where poverty is prevalent, and not all of it is retained by the central governments (some is redistributed to regional and local governments). Even so, if the governments in these two regions were to transfer all of their remaining annual mining tax receipts to the poor, it would make a small but perceptible contribution towards the zero-poverty goal in 2030. The political feasibility of such a redistribution of tax revenues is a challenge, given the historically inferior rate of income redistribution in mining economies compared with other developing economies.

Given this, the best pathway for mining to contribute to SDG 1 is via increasing industrialized mining's local demands for inputs, known as backward linkages. This is something that can be done beyond the confines of political economy constraints. Traditionally, industrialized mining has been thought to be an enclave activity, its product exported in raw form and its inputs imported from abroad. It is widely understood that mining employs relatively little unskilled labor, and that this is the reason for the poor translation of mining-led economic growth into corresponding poverty reduction. Yet mining companies have sought to increase backward linkages by sourcing more of their inputs of goods, services, and labor locally. This is exactly how the industry itself sees its contribution to SDG 1:

> Mining can help reduce poverty through direct employment, sourcing goods and services locally and the payment of taxes and royalties which enable the development of essential social and economic infrastructure. Where companies operate in remote

> areas, poverty levels can be significant. Mining's presence can either alleviate poverty through job creation and skills development, or exacerbate poverty if traditional livelihoods and ways of life are adversely affected.*

The ICMM highlights a case study of MMG's Sepon open-pit copper mine in Lao PDR, which has been operating since 2002. There, corporate monies spent on child malnutrition, water and sanitation, local procurement, microfinance, and other activities in partnership with the local and national government have increased household incomes by a factor of ten. The district around the mine is no longer amongst the 46 poorest districts in Lao. The Yanacocha gold mine in Peru has been studied both before and after a local procurement initiative, and the resultant positive impacts of that initiative on communities within 100 km of the mine are impressive. Notably, the number of households living in poverty in those communities is estimated to have dropped by 10% as a result of the initiative. In a broader study of the local impact of mining throughout Peru, populations located within the 104 districts in which there was mine production had on average a 23% rate of extreme poverty compared with a 30% rate in the 563 nearby non-producing districts. After controlling for other factors that influence poverty in a district, mining is responsible for 3 percentage points of this 7-percentage-point difference, though part of the benefit from mining appears to be due to in-migration of wealthier and more educated workers to the district. Other studies of mining in developing economies also show modest increased wealth amongst local households, though again these effects do not extend beyond the host district in which the mine is located.

Recalling that an acceleration of the natural rate of poverty reduction is required under SDG 1, and given the relatively modest poverty-reducing effects of mining to date, an increased rate of local procurement from existing mines is required by 2030 if the industry is to make a significant contribution to SDG 1. There is good news as to this possibility: in two important mining countries for which there is data, only 67% of the 55 mining companies with operations in South Africa mention local content in their public documents, and only 53% provide statistics on their local procurement. In Namibia the numbers are 60% and 27% respectively. Not only is there room for growth in local procurement for mines throughout SSA and SA, that growth is likely. The pressures for social license are a motivating factor in this regard, as are claims of lowered procurement costs. International agencies such as Deutsche Gesellschaft für Internationale Zusammenarbeit (GIZ) GmbH and Engineers without Borders Canada have initiated a Mining Local Procurement Reporting Mechanism to assist mining companies with local procurement programs. Governments themselves have local content policies in place, aided by groups such as the Canadian International Resources and Development Institute, the Intergovernmental Forum on Mining, Metals and Sustainable Development, the Columbia Center on Sustainable Investment, and the World Bank Energy and Extractives Global Practice.

The challenge will be to maintain and grow these backward linkages at a time when mining is becoming more capital intensive and more technologically

* https://www.icmm.com/en-gb/metals-and-minerals/making-a-positive-contribution/no-poverty

advanced, reducing the natural demand for local workers and local inputs. To quote the International Institute for Sustainable Development's Aaron Cosbey, "we need to start the conversation now, between governments, mining investors, between local communities, to try to find workable answers" to this challenge.

## LESSONS LEARNED

At the local level mining reduces poverty via backward linkages. Given the relatively short time frame set out by the SDGs, it is through expanding this mechanism, rather than through increased mining investment and mining-led economic growth, that the industry can best contribute towards eliminating extreme poverty. Given the geographical concentration of extreme poverty in SSA and SA, the most opportunity comes from the extensive, established mining operations in the SSA countries, and less so in SA where there is more limited mining activity.

## BIBLIOGRAPHY

African Union (2009) Africa mining vision.

Alexeev M, Conrad R (2009) The elusive curse of oil. *Review of Economics and Statistics* 91(3):586–598.

Aragón FM, Rud JP (2013) Natural resources and local economic development: Evidence from a Peruvian gold mine. *American Economic Journal: Economic Policy* 5(2):1–25.

Besley T, Burgess R (2003) Halving global poverty. *Journal of Economic Perspectives* 17(3):3–22.

Bhattacharyya S, Resosudarmo BP (2015) Growth, growth accelerations, and the poor: Lessons from Indonesia. *World Development* 66:154–165.

Bjørnland HC, Thorsrud LA (2016) Boom or gloom? Examining the Dutch disease in two-speed economies. *Economic Journal* 126(598):2219–2256.

Columbia Center on Sustainable Investment (2016) Mapping mining to the sustainable development goals: An atlas.

Cosbey A, Mann H, Maennling N, Toledano P, Geipel J, Brauch MD (2016) *Mining a Miracle? Reassessing the Shared-Value Paradigm in Light of the Technological Advances in the Mining Sector.* International Institute for Sustainable Development, Canada. ISBN 978-1-894784-75-7.

Davis GA (2009) Extractive economies, growth and the poor. In: Richards JP (ed) *Mining, Society, and a Sustainable World.* Springer–Verlag, Berlin, pp 37–60.

Davis GA (2011) The resource drag. *International Economics and Economic Policy* 8(2):155–176.

Davis GA (forthcoming) Large-sample evidence of income inequality in resource-rich nations. *Mineral Economics.* https://doi.org/10.1007/s13563-019-00207-1.

Davis GA, Vásquez Cordano AL (2013) The fate of the poor in growing mineral and energy economies. *Resources Policy* 38(2):138–151.

Deutsche Gesellschaft für Internationale Zusammenarbeit (GIZ) GmbH (2016). Linkages to the resource sector: The role of companies, government and international development cooperation.

Deutsche Gesellschaft für Internationale Zusammenarbeit (GIZ) GmbH (2017). Mining local procurement reporting mechanism.

Engineers Without Borders Canada (2017) The relationship between local procurement strategies of mining companies & their regulatory environments.

Galbraith JK, Kum H (2005) Estimating the inequality of household incomes: A statistical approach to the creation of a dense and consistent global data set. *Review of Income and Wealth* 51(1):115–143.

Gamu J, Le Billon P, Spiegel S (2015) Extractive industries and poverty: A review of recent findings and linkage mechanisms. *Extractive Industries and Society* 2(1):162–176.

Geipel J (2017) Local procurement in mining: A central component of tackling the resource curse. *Extractive Industries and Society* 2(3):434–438.

Gotz JVD, Barnwal P (2019) Mines: The local wealth effects and health effects of mineral mining in developing countries. *J Dev Econ* 139:1–16.

Highgrade Media (n.d.) Local content: New technology, new reality. http://www.highgrade.media/talks/aaron-cosbey-local-content-new-technology/

International Council on Mining and Metals (2018) ICMM members' tax contribution: Focusing on corporate income tax and royalties.

Intergovernmental Forum on Mining, Minerals, Metals and Sustainable Development (2019) Local content policies in the mining sector: Scaling up local procurement.

James AG (2015) The resource curse: A statistical mirage? *J Dev Econ* 114:55–63.

James AG (2019) Fata morganas in oil-rich, institution-poor economies. *Resour Policy* 60:234–242.

Loayza NV, Raddatz C (2010) The composition of growth matters for poverty alleviation. *J Dev Econ* 93:137–151.

Loayza N, Rigolini J (2016) The local impact of mining on poverty and inequality: Evidence from the commodity boom in Peru. *World Dev* 84:219–234.

Mamo N, Bhattacharyya S, Moradi A (2019) Intensive and extensive margins of mining and development: Evidence from Sub-Saharan Africa. *J Dev Econ* 139:28–49.

Page L, Pande R (2018) Ending global poverty: Why money isn't enough. *J Econ Perspectives* 32(4):173–200.

Ross ML (2007) How mineral-rich states can reduce inequality. In: Macartan H, Sachs JD, Stiglitz JE (eds) *Escaping the Resource Curse*. Columbia UP, New York, pp 237–255.

Segal P (2011) Resource rents, redistribution, and halving global poverty: The resource dividend. *World Dev* 39(4):475–489.

Smith B (2015) The resource curse exorcised: Evidence from a panel of countries. *J Dev Econ* 116:57–73.

United Nations (2019) World economic situation and prospects 2019.

World Bank (2015) A practical guide to increasing mining local procurement in West Africa. Report no. AUS6324.

GOAL 2

# End Hunger, Achieve Food Security and Improved Nutrition and Promote Sustainable Agriculture

## *Zero Hunger: Fertiliser Industry's Potential Contribution – A Case Study*

*P.B. Anand, C. Parra and J. Weakliam*

### CONTENTS



## INTRODUCTION

The goal of zero hunger depends on sustaining agricultural productivity. However, productivity of agricultural land is more fragile now than at any time in the last century due to the intensive demands we make on it. Certain fertilisers can be produced by processing raw materials that occur in nature and can be mined while many others are entirely synthesised. During the coming decades, mining fertiliser projects will have significant potential to play a key role in enhancing agriculture and food security globally. This chapter presents a case study of a mining project for potash in Eritrea in East Africa.

Agriculture and food security involve an important part of the population. Worldwide some 28 per cent of all workers are in agriculture but this proportion can be much higher in sub-Saharan African countries: over 92 per cent in Burundi, 81 per cent in Chad, and over 70 per cent in Niger, Central African Republic, Somalia, Malawi, Mozambique, and Uganda (World Bank, 2018). Therefore, improving the social conditions of the agrarian population must consider multidimensional and interconnected development factors such as: demographic and social factors; land ownership; access to credit and finance; agricultural extension services; farm inputs and access to markets and information and the role of public institutions to create sound policies and implement efficient programs; environmental and climate conditions. Obviously, access to information and scientific knowledge is needed to enable these farmers to address challenges related to the quality of soil, availability of fertilisers and their fundamental connections with productivity of crops and economies of scale.

## UNDERSTANDING THE COMPLEXITY OF THE FUNCTION: FERTILISER, SMALL FARMERS, AGRICULTURE AND FOOD SECURITY

For several decades, fertiliser demand has been low in sub-Saharan Africa. In a study of Burkina Faso, Koussoube and Nauges (2016) noted that:

> These findings suggest that the low uptake of chemical fertilisers might have been driven by factors other than profitability, including insufficient supply of subsidised fertilisers to farmers in need. Our results also call for increasing the availability of credit to farmers in order to encourage adoption of chemical fertilisers.

Their study highlights that while incentives and the traditional explanation of relative cost of fertiliser to crop are important, other relevant factors include delivery of the fertiliser to the farm gate and education and support programmes, including credit and other social support mechanisms.

In another study, Bruilhe and Barriero-Hurle (2012) review the experience of fertiliser subsidies in 14 African countries and conclude that:

> The new generation of input subsidies ('smart' subsidies) brings innovations in design (e.g. targeting; vouchers) to support both the most constrained farmers and encourage the development of input markets. Available evidence, albeit very limited, suggests that such programmes have been effective in raising fertiliser use, average yields and agricultural production but that their success is highly dependent on implementation.

Their study highlights the importance of implementation of such Smart subsidies.

From a study using Living Standards Measurement Studies Data for six countries and covering a very large sample of over 22,000 households, Sheahan and Barrett (2017) shed light on several preconceptions about African smallholder agriculture. They find that contrary to the established stylised fact that African farmers use very small amounts of inorganic fertilisers, the LSMS study showed that actually there is quite a range and some farmers do use significant amounts of fertiliser. What is interesting, however, was that irrespective of whether soil quality is good, mediocre or poor, farmers seemed to be using about the same quantity of fertiliser and that very few farmers used credit to purchase fertilisers. Another important finding in this study concerns the country effect:

> Quite surprisingly, the overwhelming amount of variation, indeed nearly half (45%), is accounted for by the country dummy variables. Even controlling for a wide range of important observable household-level and agro-ecological variables, some combination of other policy, institutional, or macroeconomic variables explain most of the micro-scale variation in inorganic fertilizer use in this unprecedentedly large sample of over 22,000 households. … This is a significant finding, as clearly the policy and operating environments facilitated by governments matter.

Potash and phosphate are among fertilisers that occur in certain mines in nature. Potash of good quality can help plant growth and the robustness of plants to withstand weight, wind and other weather conditions. Much potash supply presently comes from Canada, Belarus, Russia and China. Canada produces some 22 million tonnes of potash (almost a third of global production). However, for many farmers in Africa it is far too expensive and therefore out of reach. The project currently in progress in Eritrea has the potential to significantly change and improve current availability of fertiliser and generate positive effects on crop productivity, especially in Africa.

## COLLULI PRODUCTION: POTASH AND OTHER MINERAL FERTILISERS

Potash is the common term for fertiliser forms of the element potassium (K). The name derives from the collection of wood ash in metal pots when the beneficial fertiliser properties of this material were first recognised many centuries ago. Potassium is one of three key fertiliser 'macro-nutrients' essential for healthy soil and plant growth. It is generally used in combination with the other two macronutrients, nitrogen and phosphorus, to produce a range of fertilisers, the type used being dependent on the soil to which it will be applied.

Potassium is essential to the workings of every living plant cell. It not only plays an important role in a plant's use of water but also helps regulate the rate of photosynthesis. Potassium promotes the growth of strong stalks, protects plants from extreme temperatures and enhances their ability to cope with stress. Importantly, there is no substitute for potash.

Commonly, potash refers to potassium chloride or muriate of potash (MOP); however, a number of other potash variants exist with premium potash types containing micronutrients as well as macronutrients such as potassium sulphate, potassium magnesium sulphate and potassium nitrate.

Potassium-bearing minerals are typically found in areas where ancient inland seas have evaporated, leaving behind their minerals. The Danakil salt basin is a rare example, as it is one of only very few known potash deposits where kainite, a sulphate-bearing mineral key for potassium sulphate production, is found in solid form.

The potassium-bearing salts within the Colluli resource have the unique capability of producing a diverse range of potash types including sulphate of potash (SOP or potassium sulphate), muriate of potash (MOP or potassium chloride) and sulphate of potash magnesia (SOPM or potassium magnesium sulphate).

## PRODUCTIVITY IN AGRICULTURE IN ERITREA: A MACRO PERSPECTIVE

It is estimated that around 1.5 million of people in Eritrea work directly in the agriculture sector in the country, representing around 60 per cent of the country's total labour force. Because of that, around 80 to 90 per cent of the population has a direct connection with the sector, evidencing how important this economic activity is for the national sustainable development agenda, and how crucial it is to understand the relationship between fertiliser and productivity in agriculture.

At macro level, productivity in agriculture in Eritrea can be characterised by assessing the relationship between employment in agriculture and the share of agriculture in Eritrean gross national income (GNI); in other words, by assessing the capacity of the Eritrean labour force to generate value-added in this sector. The current capacity of the labour force to generate value-added in agriculture is 17 per cent and this has been declining slightly. It is highly probable that the agricultural sector is experiencing well-known symptoms of structural constraint linked to low levels of productivity. Though Eritrea has good export potential there is also significant domestic demand and the infrastructure constraints make exports uncompetitive. Data from other secondary sources and our primary interviews are in line with these observations.

## POTASH AND MINERAL FERTILISER BEYOND ERITREA: FERTILISERS GALORE IN AFRICA

Productivity, soil conditions and climatic factors remain as a challenge and a pressure on agriculture for many countries in Africa (Figure 2.1). Notwithstanding this, African productivity has been progressing consistently beyond the specific relationship between employment and GDP in agriculture and taking into account a more multidimensional approach.

Despite the fact that most African nations have some of the lowest fertiliser consumption per hectare compared to farmers elsewhere, interesting, large fertiliser investments are in progress. Morocco already boasts one of the world's largest

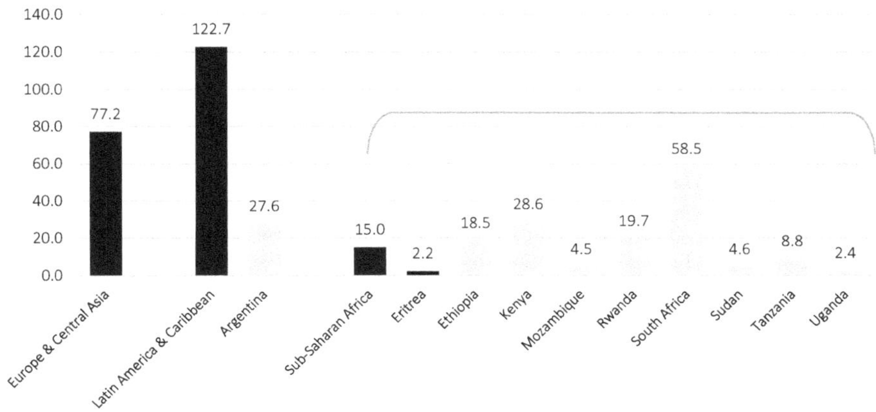


**FIGURE 2.1** Fertiliser consumption (selected countries) 2015 (kilograms per hectare of arable land).

phosphate mines and the OCP Group reports that these contribute to 20 per cent of Morocco's exports. A news report in *The East African* (April 2018) mentioned that Tanzania Petroleum Development Corporation is to build a $3 billion fertiliser factory in Lindi and Mtwara to utilise its natural gas to produce 3.8 million tonnes of fertiliser per annum. *The African Courier* reported in July 2017 that Nigeria had commissioned the world's biggest single-train urea plant in Port Harcourt. The report says that the plant built by Indorama Eleme Fertilisers and Chemicals Limited has a production capacity of 1.5 million tonnes of urea per annum.

In this context, the Colluli project in Eritrea has a unique potential opportunity to generate significant change. When the project reaches its full production capacity it will become the world's largest SOP producer.

The natural advantages of Colluli (its mineral is available close to the surface and it is located within 100 kilometres of the Red Sea, simplifying connection to international markets) can boost Eritrean and African agriculture, employment and GDP. Potash and other agrochemicals from Colluli combined with infrastructure and capacity-building can improve productivity, agricultural output and food security as well as create further employment in these countries and contribute to boosting rural incomes and reducing poverty.

## POTASH, POTATO AND POTENTIAL FOR PROSPERITY IN ERITREA

In 2017, recognising the importance of productivity and research in agriculture, the EU agreed to fund a major agricultural development initiative in Eritrea under the Eritrea Ireland Development partnership. The implementation of this project is led by Irish development agency VITA in partnership with Teagasc, UCC, UCD, LUKE (Finland), Self Help Africa (SHA), Hamelmalo Agricultural College, (HAC), Eritrea, Ministry of Agriculture Eritrea (MoA) and National Agricultural Research Institute (NARI), Eritrea. The focus will be on three key objectives: (i) new innovations on smallholder farms with a focus on the contribution of dairy livestock;

(ii) strengthening of research, education and innovation support capacity in Eritrea; (iii) new initiatives in relation to youth employment.

## COLLULI POTASH PRODUCTION, AS AN EXTERNAL FACTOR, IS A CRITICAL COMPLEMENT TO THE FINAL OBJECTIVES OF THIS PROJECT.

Potash can benefit many food crops in Eritrea as well as in other parts of Africa. In recent years the Eritrean Government has come to recognise the potato as a strategic crop which can enhance national food security. Current potato production is 30,000 tons annually but the Ministry of Agriculture is targeting an increase to 150,000 tons annually by 2030, which would make the potato the single most important growth crop in Eritrea. Such production would provide substantial food security, incomes to farmers and affordable staple food to consumers, and would drive food exports and processing.

The special value of the potato in Eritrea relates to its productivity per scarce resource (land area, irrigation water, carbon emissions). Recent studies indicate that application of sulphate of potash (such as will be produced in Colluli) increases productivity per each of these scarce resource and especially for water, which is the critical constraint in Eritrea. Studies and soil analysis conducted in Teagasc in Ireland and Hamelmalo University in Eritrea show that application of sulphate of potash provides Eritrean farmers with yield increases of 15 to 40 per cent, and averaging 20 per cent. On the other hand, without regular application of potash, soil fertility and potato yields will diminish over time.

In terms of SDG 2 and delivering increased agriculture productivity, the potato has unique potential as a supplement to cereal crops. While current yields average 12 tons, a pilot programme in 2016/17 reveals that farmers are sustaining yields in excess of 36 tons where they use improved clean seed and apply fertiliser.

Taking the target of 150,000 tons of potato production by 2030 and assuming a contribution by potash of 20 per cent to sustainable yields, would indicate that application by 10,000 farmers of sulphate of potash can contribute 20 per cent towards the 150,000 or 30,000 tons annually. This implies the potential contribution of potash to potato production as equivalent to almost 10 per cent of the current level of national food production.

The potato can be a part of the diversification and resilience strategy in terms of its resilience to climatic conditions as well as to certain pests etc. It can also contribute to nitrogen fixation and thus be part of a crop rotation system, contributing to the increased productivity of other cereal crops. Also, given the location of Eritrea, there is good export potential to the Middle East.

## FINAL REFLECTION AND POTENTIAL CONTRIBUTION TO SDGS

Considering the characteristics of employment and productivity in agriculture in Eritrea and SDG 2 (End hunger, achieve food security and improved nutrition and promote sustainable agriculture), the production of Colluli could improve agricultural productivity and the incomes of small-scale food producers such as family farmers. In addition, the introduction in the country of a high-quality fertiliser can create a new baseline for maintaining the genetic diversity of seeds, cultivated plants and farmed and domesticated animals and their related wild species.

Considering this potential, it is crucial to consolidate different alliances and partnership efforts – such as the Eritrea-Ireland Development Partnership for Agriculture & Food Security in Eritrea (implemented by VITA in partnership with Eritrean Ministry of Agriculture and the Irish State Agriculture Agency, Teagasc) – as catalyst mechanisms for agriculture development; thus enabling specific effects on poverty alleviation, food security and socioeconomic and sustainable issues linked to agriculture such as productivity and the social conditions of small farmers in Eritrea.

Availability of fertilisers must be considered as an extremely important first step in the development pathway of Eritrea. However, it will be necessary to have intensive planning for training in the use of fertilisers, developing logistics and fertiliser distribution mechanisms, research into crops, infrastructure development including for irrigation and storage, market analysis, among other key variables.

Progress in terms of agriculture and food security in Eritrea (and by extension in Africa) represents a major challenge that necessarily should include a collaborative research component in order to capitalize and accumulate knowledge and capacities within Eritrea.

The country could use experience from Colluli to:

- Strengthen internal capacities to define problems.
- Set objectives and priorities.
- Conduct sound scientific programmes.
- Build sustainable institutions, including through community-based water management institutions and agricultural co-operatives.
- Identify solutions to key specific national problems.

Considering this necessity, one of our recommendations would be to consider the creation of a research centre for agriculture, food security and agricultural finance in Eritrea and Africa to develop internal capacities and capitalise the opportunities of Colluli to:

- Accumulate and organise a baseline of knowledge to support future action and programs in relation to agriculture and food security.
- Evaluate technical assistance and technology transfer to build national capacity in Eritrea in agriculture and food security.
- Assist national authorities in Eritrea to develop national strategies.
- Support research in relation to soil, productivity, seed, etc.
- Develop excellent market research and financial analytical capacity.

We are aware that promoting research and innovation is a highly capital-intensive activity and a small country such as Eritrea has many disadvantages. However,

Colluli offers the potential for Eritrea to divert some of the windfall gains from the mining project to develop a world-class research centre in collaboration with international organisations such as Teagasc, FAO, CGIAR and IFDC and other regional agricultural research centres in Africa.

## REFERENCES

Bruilhe Z and Barreiro-Hurle, J (2012) *Fertiliser subsidies in sub-Saharan Africa, ESA Working paper number 12-04*, Rome: Food and Agriculture Organisation.

Government of the State of Eritrea (2010) *Eritrea Population and Health Survey 2010,* Oslo: Department of Statistics and Fafo Institute for International Studies.

Koussoube, E and Nauges, C (2016) Returns to fertiliser use: Does it pay enough? Some new evidence from Sub-Saharan Africa. *European Review of Agricultural Economics*, 44(2), 183–210.

Sharma, D and Bhatnagar, P (2015) Corporate social responsibility of mining industries. *International Journal of Law and Management*, 57(5), 367–372. doi:10.1108/IJLMA-03-2014-0022.

Sheahan, M and Barrett, C (2017) Ten striking facts about agricultural input use in Sub-Saharan Africa. *Food Policy*, 67, 12–25.

World Bank (2018) World development indicators data.

## ADDITIONAL RELEVANT RESOURCES

African Development Bank (2018) *African Economic Outlook 2018,* Abidjan: ADB.

Brew, Y, Junwu, C and Addae-Boateng, S (2015) Corporate social responsibility activities of mining companies: The views of the local communities in Ghana. *American Journal of International Business Management*, 5, 457–465.

Danakali Ltd (2016) *Definitive feasibility study-economic outcomes*, Perth: Danakali.

Gilberthorpe, E, Agol, D and Gegg, T (2016) 'Sustainable mining'? Corporate social responsibility, migration and livelihood choices in Zambia. *The Journal of Development Studies*, 52(11), 1517–1532.

Global Reporting Initiative (2015) Sustainability reporting guidelines and metals and minerals sector supplement.

Global Reporting Initiative (2018) G4: Global reporting initiative guideline 4.

Global Reporting Initiative (2018) Mining and Metals, G4 Sector Disclosure.

Government of Canada (2018) *Doing business the Canadian way: A strategy to advance corporate social responsibility in Canada's extractive sector abroad*, Ottawa: Government of Canada.

ICMM (2016) Making a positive contribution to the SDGs-ICMM Guide.

Martinik Bosch Sell Pty Ltd (2016) Draft environmental and social impact assessment, report prepared for Colluli mining share company (three volumes).

Rodrik, D (2007) *One economics, many recipes: Globalisation, institutions and economic growth*, Princeton, NJ: Princeton University Press.

Siyobi, B (2015) Corporate social responsibility in South Africa's mining industry: An assessment. *Policy Briefing 142, Governance of Africa's Resources Programme, SAIIA*.

UNDP (2016) *Mapping mining to the sustainable development goals: An Atlas*, Geneva: World Economic Forum.

Wilson, DL et al (2016) Avoided emissions of a fuel-efficient biomass cookstove dwarf embodied emissions. *Development Engineering*, *1*, 45–52.

# GOAL 3 Ensure Healthy Lives and Promote Well-Being for All at All Ages

*Francesca Viliani and Mirko S. Winkler*

## CONTENTS



## SUMMARY

### Key Stakeholders

Disease prevention and promotion of good health in the workforce and project-affected communities can be achieved only if a diverse group of stakeholders are considered and involved at different levels:

- Individual level – the worker;
- Social level – project-affected communities and workers' dependents;
- Economic level – the company and its shareholders;
- Governance level:

- Project area – local authorities, local health system, other operating companies, service providers, civil society; and
- National/global level – national authorities, industry associations, financial sector and investors, UN agencies, universities and other eventual actors.

## Direct Mining Contribution/Drivers

For many years, mining operations have been considered to be among the most hazardous workplaces (Cho 1978), but investment in safety has led to a reduction in fatalities over the past decades. For example, members of the International Council on Mining and Metals (ICMM) reported 50 fatalities[1] globally in 2018 (ICMM 2018). In addition to accidents and injuries, a significant number of mine workers have been affected by disabling occupational diseases such as hearing loss, pneumoconiosis,[2] and the effects of vibration,[3] among others. Despite awareness of potential health risks associated with mining, companies have not always recognised the importance of occupational health, and even less the role of public health, to support sustainable projects.

In this context, the most important direct contribution of a mining project should be the adoption of a comprehensive health management system (e.g. in line with the Global Reporting Initiative (GRI) Standard 403: Occupational Health and Safety 2018 (GRI 2018)), coupled with the mining industry's commitment to community engagement and investments. For mining to contribute to health and well-being, each project should avoid generating negative health impacts on workers and communities, along with and enhancing health benefits (IFC 2012). This aligns with Principle 5 of ICMM that requires companies to continually improve health and safety performance, with the ultimate goal of zero harm. For example, Anglo American (Anglo American 2019) has committed to a goal of zero harm "by leading with safety and investing in health and well-being" as well as aiming to achieve the SDG targets for health in all their host communities.

## Contribution to SDGs

SDG 3 aims to ensure healthy lives and promote well-being for all at all ages. Health and well-being are a prerequisite for sustainable development, as well as a result of it. Thus, they are central to the achievement of the 2030 Agenda for Sustainable Development.

But what does good health actually mean? According to the World Health Organization (WHO), health is "a state of complete physical, mental and social well-being and not merely the absence of disease or infirmity" (WHO 1946). The health status of individuals and populations is, thus, determined by complex interactions between various personal, social, economic, political, and environmental factors. SDG 3 is therefore closely interlinked with all the other SDGs (Figure 3.1). There is plenty of evidence connecting, for example, SDG 1: "no poverty", and SDG 10: "reduced inequalities with health gains" (Commission on Social Determinants of Health 2008). Moreover, SDG 4: "quality education" (OECD 2006), SDG 5: "gender

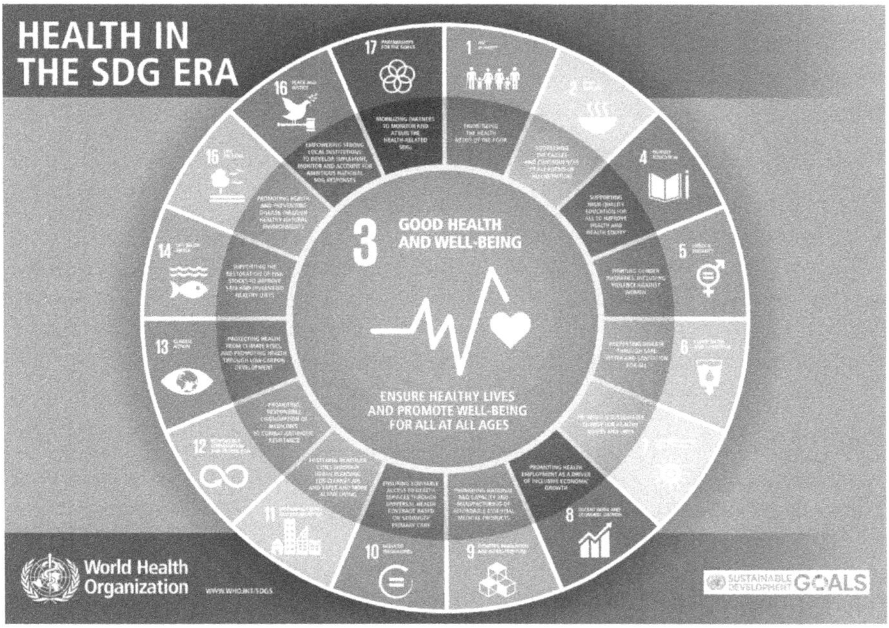


**FIGURE 3.1** The interconnectedness of good health and well-being with the other SDGs. (Source: https://www.who.int/sdg/en/)

equality" (Connel 2012), and SDG 6: "clean water and sanitation" (WHO 2019) are all examples of key determinants of health that directly influence the well-being of people and population groups.

## Key Regulations and Public Policies

Several national, international, and financial regulations exist regarding occupational health and safety (ILO 2014) and health impact assessment (HIA) (Winkler 2013). Regulatory frameworks for environmental impact assessment (EIA) are also considered important public policies supporting SDG 3, as they commonly embrace environmental and other determinants of health. In some countries, sector-specific or disease-specific regulations exist (e.g. on HIV in the workplace and non-discrimination).

## Corporate Policies and Private Programs

Health and well-being should be an integral component of the management system of all mining operations. Furthermore, as the mining sector can generate health impacts on workers and communities (Stephens 2002), such systems should not rely exclusively on occupational health and safety system and practitioners (ICMM 2012) but should include a range of disciplines in order to fully consider the wider cumulative and social-ecological dimensions of health.

## ANALYSES OF MINING CONTRIBUTION TO SDGS

SDG 3 "good health and well-being" is articulated through 13 targets, supported by 26 indicators (https://sustainabledevelopment.un.org/sdg3).

The following section introduces the different health targets relevant to mining and describes the pathways from initial resource extraction to ongoing production that can lead to population health impacts. For each health target, both the risks to organisations and the opportunities available are considered.

### Health Target 3.3

*End the epidemics of AIDS, tuberculosis, malaria, and neglected tropical diseases, and combat hepatitis, water-borne diseases, and other communicable diseases.*

This target brings together all the infectious diseases[4] that are transmitted between humans either directly (from person to person) or via vector.[5] Some of these diseases, mainly HIV/AIDS, TB, and malaria, are among the major health problems faced by the industry in certain geographical settings. ICMM developed guidance for how to deal with these diseases (ICMM 2008).

If companies are not proactive, the development and operation of mining projects may further exacerbate the existing burden of communicable diseases or the spread of diseases that previously did not exist in the impacted area. The underlying risk factors are very context- and project-specific. For example, workers may be exposed to infectious diseases at the workplace or in accommodation areas. Non-local workers may have a different epidemiological profile than the local workers and therefore a different susceptibility to infectious disease. Simultaneously non-local workers may be carrying infectious diseases and transmit these to co-workers, as well as local communities. In addition, the transmission of infectious diseases can be altered through changing behaviour in the local population and impacts on ecosystems as summarised in Table 3.1.

The opportunity, however, is for mining companies to reduce the burden of communicable diseases in their workforce and project-affected communities, especially in low- and middle-income countries (LMICs) in the global south. For example, the mining sector was the first sector to respond to the HIV/AIDS epidemic in southern Africa (IFC 2004).

Another well-known example is the AngloGold Ashanti mining project in Ghana. When the management realised that the economic burden of malaria on their operations was becoming increasingly significant, they decided to support a comprehensive malaria program, targeting not just workers but also surrounding communities. The integrated malaria program went on to receive financial support from the Global Fund and became a showcase of a successful public–private partnership in the mining sector (Lambe 2018; ADB 2016).

**TABLE 3.1**
**Diseases in SDG 3, Target 3.3 and Its Main Drivers**

| Disease clusters | Main drivers |
|---|---|
| *Sexually transmitted infections (STIs), including HIV* | Predominantly male labour force migrating to the mining areas and living in non-family working camps; existing poverty and inflation could lead to commercial sex. |
| *Vector-related and zoonotic diseases* | The project alters the local environment and supporting ecosystems, and this may influence the breeding capacity and presence of vectors and animals; migration of workers and other population groups can introduce new infections to the local area. |
| *Water-borne and water-washed diseases* | Possible competition over resources such as water between the project and local communities; in-migration in the area puts additional pressure on existing sanitation and water infrastructures; possible pollution and contamination of drinking, irrigation, and use-water sources; poor hygiene practice. |
| *Tuberculosis* | Overcrowding, which can be aggravated by inflation and poverty (important to ensure enough accommodation space for workers (IFC and EBRD 2009)); in-migration of workers or camp followers from areas/countries with high TB and/or HIV prevalence. |

Companies can also leverage their health capabilities to support research to tackle diseases which can have profound and lasting local, national or even international impact. An International SOS clinical team working for Newcrest, a mining company operating in Papua New Guinea, discovered that azithromycin (an affordable and easy-to-administer antibiotic) cures the tropical disease yaws (Mitja 2015). This initiative drove the global WHO strategy to decide to eradicate the disease worldwide by 2030, directly contributing to the achievement of this target.

Many more examples of the link between SDGs and mining exist from projects that partnered with local, provincial, and national organizations to provide clean water and sanitation to community members in their concession area as a way to reduce diarrhoeal diseases (GBC 2013).

## Health Target 3.4

*Reduce premature mortality from non-communicable diseases through prevention and treatment and promote mental health and well-being.*

## Health Target 3.5

*Strengthen prevention and treatment of substance abuse and harmful use of alcohol.*

## Health Target 3.9

*Substantially reduce deaths from hazardous chemicals and air, water, and soil pollution and contamination.*

Health targets 3.4, 3.5 and 3.9 bring together all chronic, non-communicable diseases (NCD) and their associated risk factors. These diseases tend to be of long duration, in some instances chronic, and are the result of a combination of genetic, physiological, environmental, and behavioural factors. The main types of NCDs are cardiovascular diseases (e.g. heart attacks and stroke), cancers, chronic respiratory diseases (e.g. chronic obstructive pulmonary disease and asthma), diabetes, and mental illness.

There are several risks for mining projects to consider with respect to NCD in workers, communities, as well as health systems. In many cases, NCDs are already the largest health burden for workers, in terms of both morbidity and mortality; in most countries, the incidence of these diseases is rising, and increasingly so in LMICs, where the majority of mining projects exist. Furthermore, mining workers may face elevated metabolic, behavioural, and occupational risk factors for NCD (Chalmers 2012; Rodriguez- Fernandez 2015; Mawaw 2017). This increases not only the burden of disease in the workers but also the health costs associated with managing these chronic conditions. Once a mining project closes or a worker retires, these costs are transferred to the local health systems.

Local communities are a source of labour and may present with a high burden of NCD. In many rural areas of LMICs, the burden of NCD might actually be under-reported by the health authorities, as the local health systems might not yet have rolled out full capacities to diagnose and manage NCDs and health information systems might not be robust enough to track the underlying burden. The exposure to any form of environmental pollution will further increase the burden of NCD (Pruss Ustun 2019). This ranges from an increased risk of cancer and respiratory problems, to promoting birth defects. Health impacts associated with environmental pollution are of particular concern in communities that are directly exposed to mining operations. Last but not least, the changes in lifestyle and economic development brought about by mining projects are associated with a range of behavioural risk factors that promote the development of NCDs, such as lack of physical exercise, unhealthy dietary practices, or alcohol abuse.

Mental health issues are another critical health concern to be considered in the workforce (Ling 2016), as well as workers' families and affected communities at large (Hossain 2013). Mental health issues such as depression and anxiety disorders have been reported in contexts that undergo rapid changes such as boom-and-bust cycles (Shandro 2011). The mining environment was shown to be conducive to substance abuse and mental illness for several reasons: a male-dominated environment with no family accommodation, split lifestyles further exacerbated by a fly-in fly-out (FIFO) working pattern, long working hours organised around shifts, and the often remote location of these operations in harsh climatic conditions. Lack of access to family and friends, social supports, coupled with often problematic telecommunications in remote settings, compounds this risk.

Many companies have in place policies that aim to reduce the safety risks associated with substance abuse. However, substance abuse is often not an isolated problem. Therefore, a holistic health-promoting environment that comprises awareness programmes, healthy working conditions, and appropriate accommodation arrangements holds the potential for lasting and comprehensive solutions. For example, new guidance promotes mentally healthy workplaces and encourages companies to consider the mental health consequences of their operations (Commission for Occupational Safety and Health 2019).

Mining companies are becoming increasingly aware of the challenges posed by NCDs on their operations, and are therefore addressing associated risk factors more proactively. For example, PT Freeport Indonesia has supported a comprehensive tobacco control program aligned with WHO and Indonesian recommendations: tobacco use prevalence rates in their workforce have decreased significantly from baseline through five-year follow-up (Rodriguez-Fernandez 2016). This is even more impressive when considering that Indonesia's national smoking rates have risen during this time.

When designing the health management system at the project level, all the above-mentioned risk factors should be considered. Indeed, according to GRI 403, both health promotion interventions and policies targeting modifiable lifestyle patterns and environmental exposures are needed to mitigate associated health risks. Simultaneously, the rapid changes associated with mining development can offer a unique opportunity to promote a healthier lifestyle in both workers and communities. This can, however, only be achieved through a concerted effort among multiple partners done in coordination with local and national authorities.

## Health Target 3.6

*Halve the number of global deaths and injuries from road traffic accidents.*

This target deals with the worldwide challenge of road safety, but it has also an equality dimension as most fatalities on the roads occur in LMICs, disproportionately affecting vulnerable population segments such as children and young people (WHO 2018).

There are several risks for mining projects associated with road safety, depending on the location relative to the project. Controls targeting different users are shown in Table 3.2.

**TABLE 3.2**
**Road Traffic Accidents in Mining: Company Controls and Risk Factors**

| Type of road | Users | Company control and measures | Most at risk |
|---|---|---|---|
| Within the mining project | Company employees and authorised personnel | Total control over traffic rules (including speed limits and access) | Workers and light vehicles |
| Within the larger mining concession | Mix of users | Speed limits imposed; company has some traffic control | School-age children and vulnerable road users |
| Outside the mining concession | All users | Company has control exclusively over its own vehicles | School-age children and vulnerable road users |

Usually, companies put in place mitigation measures such as a traffic and transportation safety management plan. Examples of specific measures include compulsory driving examinations for all company drivers (and those of their subcontractors), vehicle maintenance, enforced speed limits, alcohol and drug tests, installing technology that monitors driver fatigue, or building school bus stops on public roads to protect children. However, the focus is in general still very much on behavioural changes of the driver/user, neglecting the great opportunities that come with safer vehicles and safer roads. These opportunities can be achieved exclusively through systemic change and partnerships among multiple actors. For example, risks for pedestrians, cyclists, and motorcyclists due to lack of specific road infrastructure can be systematically identified through HIA, for subsequently promoting targeted interventions (e.g. upgrading of high-risk roads and providing access to safe public transport; purchasing vehicles with electronic stability control; supporting effective care for the injured through provision of prompt transport to appropriate emergency and surgical care in hospital; and early access to rehabilitation services (WHO 2018).

## Health Target 3.D

*Strengthen capacity of all countries, in particular developing countries, for early warning, risk reduction, and management of national and global health risks.*

This target deals with emerging infectious diseases (EIDs)[6] and other health problems covered by the International Health Regulations (IHR) 2005 (WHO 2005). As was evidenced during the 2014–2015 Ebola outbreaks in West Africa, the mining sector is not immune to these risks, and it can play a pivotal role in either increasing the risk or enhancing controls (Viliani 2017). At the time of writing this book chapter, the covid-19 pandemic is clearly demonstrating the many threats posed by EIDs spreading across the globe, with dramatic effects at the global and local level, including the complete disruption of mining operations in many places.

Mining activities generate significant environmental, social, demographic, health, and economic changes – all of which are critical drivers for disease emergence and outbreaks. Therefore, companies should always do a comprehensive assessment of the potential risks associated with their investments from the early phase of project conceptualisation (USAID 2012a) and throughout the life of the mine (USAID 2012b). For example, the Ebola virus has been found in bats in Liberia (Columbia University Mailman School of Public Health 2019) which means that any eco-social change affecting the life patterns of these animals could increase the risk of transmission of the virus to humans, including those working on surrounding project sites.

At the same time, all disease control-and-prevention strategies implemented by mining companies, including those for emerging infectious diseases, benefit not only the organisation but the wider community in which they operate. Furthermore, emergency response plans and pandemic preparedness strategies developed in coordination with health authorities and other actors could strengthen local responses to health emergencies.

### Other Health Targets

Health targets under SDG 3 that were not discussed in the previous sub-chapters are: 3.1 – maternal mortality; 3.2 – newborns' and children's health; 3.7 – sexual and reproductive health-care services; 3.8 – universal health coverage for all; and 3.C – health workforce in developing countries. These targets encompass health problems of particular, but not exclusive, concern for developing countries. Moreover, these targets are strongly associated with the promotion and delivery of health care services and, thus, under the direct responsibility of national and local health systems.

Mining projects exert an effect on these targets mainly by impacting, and interacting with, local health systems. On the one hand, project-induced in-migration and potential project-related negative health impacts may put additional strains on the capacity of affected health systems. Consequently, progress towards the abovementioned health targets is at risk. On the other hand, mining operations might exert a positive influence on health system-related SDG 3 targets through corporate social responsibility initiatives (ICMM 2013) that promote gender equality (Knoblauch 2018), reduce inequities in affected populations (placing particular emphasis on vulnerable groups), strengthen health care services, and promote access to health care for all.

## LESSONS LEARNED

- Worker and community health and safety risks should be systematically considered by carrying out recognised processes such as occupational health risk assessments and HIA;
- A comprehensive health and safety management plan that considers the role and responsibilities of different stakeholders in dealing with identified health and safety risks, including surveillance, is essential;
- Health programs and activities implemented must adhere to and support national health policies and plans;
- Companies should support measures to address existing health problems/needs in a given community, placing particular emphasis on vulnerable groups; and
- Mining companies could contribute to long-term community well-being by establishing partnerships with all their stakeholders, including community representatives and local and national health authorities.

## CASE ANALYSIS

### NON-COMMUNICABLE DISEASES TARGET 3.4: THE CANARY IN THE GOLDMINE

*Papua, Indonesia*

**Rodrigo Rodriguez-Fernandez**

When we think of chronic or non-communicable diseases (NCDs) such as heart attacks, stroke, cancer, mental health, and obesity (NCD risk factor), we think of office-based, sedentary, affluent Western populations. However, increasingly the world has seen a rapid rise in NCD and associated risk factors in low- and middle-income countries (LMIC) where many mining operations occur. Of 56.9 million global deaths in 2016, 40.5 million, or 71%, were due to NCDs.[7] In 2016, over three-quarters of NCD deaths – 31.5 million – occurred in LMIC with about 46% of deaths occurring before the age of 70. These trends have also impacted South East Asia. In fact, the World Health Organization (WHO) estimates that over 62% of deaths within South East Asia are caused by NCDs. That is more than HIV, TB, malaria, maternal and child, and injury-related deaths combined (38%). Similarly, the number of overweight and obese individuals living in Asian countries such as China (300 million), India (150 million), and Indonesia (41 million) have surpassed Americans – USA (160 million), Mexico (52 million), Brazil (74 million) – in the last decade.

The extension of the global epidemic has also made its way to one of the most isolated sites in the entire globe – the world's largest copper and gold mine, located at 4,200 meters above sea level (asl) in the mountainous western-central highlands of Papua Province, part of the central cordillera that stretches along much of the length of the island of New Guinea. The open pit mine (Grasberg), which employs over 30,000 mostly Indonesian nationals (98%), sits within a few kilometres of and underneath Mt. Puncak Jaya (Carstensz Pyramid) and associated mountain glaciers. At 4884m asl (16024 ft), Puncak Jaya represents the highest location in the Pacific Basin between the Himalayas and the Andes. Individuals both native to the Papuan province and those who have migrated from other central areas of Indonesia who have historically been exposed to local diets consisting of locally sourced vegetables and protein are now increasingly exposed to Western diets as well as alcohol and tobacco advertising.

PT Freeport Indonesia (PTFI), an affiliate of FreePort-McMoRan, a global mining company, commissioned a study to determine the potential risk employees faced in terms of NCDs. Findings, obtained from the large worksite cohort, highlighted high and upward-trending rates of NCD risk, and provide evidence of a potential relationship between mining-community residence and certain NCD risk factors. Against this backdrop, PTFI rolled out a site-wide multicomponent prevention model which included adopting five out of the six tobacco control measures identified by WHO as well as a baseline sodium study to determine intake levels for raised blood pressure prevention. All form part of the ongoing Cardiovascular Outcomes in a Papuan Population and Estimation of Risk (COPPER)[8, 9, 10,11] study to improve heart health in the employee mining population. Primary and secondary prevention as well as control strategies in the workplace are needed to change unhealthy lifestyle habits such as poor diet and tobacco use towards reducing heavy burdens of metabolic risk factors among mine employees. The mining industry provides a captive population that may be leveraged to influence the health and well-being of employees as evidence by well-designed workplace programs.[12]

## NOTES

1. Fatality frequency rate per 1 million hours worked: 0.022 (calculated by dividing the total number of fatalities by the hours worked at the companies, and then multiplied by 1 million).
2. A disease of the lungs due to inhalation of dust, characterized by inflammation, coughing, and fibrosis. The most common workplace mineral dusts that are known to cause pneumoconiosis are asbestos, silica (rock and sand dust), and coal dust.
3. Whole Body Vibration (WBV) caused by poorly designed or poorly maintained vehicles, platforms, or machinery, causing or exacerbating other health effects such as lower back pain (damage to vertebrae and discs, ligaments loosened from shaking), motion sickness, or bone damage.
4. "Infectious diseases are caused by pathogenic microorganisms, such as bacteria, viruses, parasites or fungi; the diseases can be spread, directly or indirectly, from one person to another. Zoonotic diseases are infectious diseases of animals that can cause disease when transmitted to humans."
5. An insect or a living carrier that transports an infectious agent from an infected individual or its wastes to a susceptible individual or its food or immediate surroundings.
6. EIDs are diseases that have either appeared in a population for the first time, or that may have existed previously but are rapidly increasing in frequency or geographic range or are old infections re-emerging as result of antimicrobial resistance.
7. Non-Communicable Diseases. World Health Organisation. Geneva, Switzerland 04/11/2019 https://www.who.int/gho/ncd/mortality_morbidity/en/
8. Rodriguez-Fernandez R, Farrand C, Cridland Raspail L, Viliani F, Amiya RM. Goliath versus Goliath: leveraging private mining industry resources for public health gains in tobacco control. Global Heart 2016;11(Suppl):e21–e22.
9. Rodriguez-Fernandez R, Rahajeng E, Viliani F et al. Noncommunicable disease risk factors in Indonesian miners: longitudinal findings from the Cardiovascular Outcomes in a Papuan Population and Estimation of Risk (COPPER) Study. Occup Environ Med 2015:1–8. doi:10.1136/oemed-2014–10266
10. Rodriguez-Fernandez R, Amiya RM et al. The double burden of disease among mining workers in Papua, Indonesia: at the crossroads between Old and New health paradigms. BMC Public Health 2016;16:951.
11. Rodriguez-Fernandez R, Farrand C, Kusuma P, Webster J. Estimation of salt intake by 24-hour urinary sodium excretion: findings from the Cardiovascular Outcomes in a Papuan Population and Estimation of Risk (Copper) Study. Global Heart 2016;11(Supplement):e19.
12. Rodriguez-Fernández R. Governments must catch up with workplace non-communicable disease prevention May 2018. Occupational Medicine 68(3):158–159 DOI: 10.1093/occmed/kqy041.

## REFERENCES

African Development Bank (ADB). 2016. *AGA Malaria and Public-Private Partnerships in Ghana's Health Sector.*

Anglo American 2019. *Sustainability.* Retrieved from: https://www.angloamerican.com/sustainability.

Chalmers L, O'Neill E. 2012. Diabetes - A new challenge for the mining industry. *Paper presented at Queensland Mining Industry Health and Safety Conference.* Retrieved from: http://www.qldminingsafety.org.au/_dbase_upl/Diabeties%20a%20new%20challenge%20for%20the%20mining%20industry.pdf.

Cho KS, Lee SH. 1978. Occupational health hazards of mine workers. *Bulletin of the World Health Organization* 56(2):205–218. Retrieved from: https://apps.who.int/iris/handle/10665/261731.

Columbia University Mailman School of Public Health. 2019. Scientists discover Ebola virus in West African bat. Retrieved from: https://www.eurekalert.org/pub_releases/2019-01/cums-sde012319.php.

Commission for Occupational Safety and Health. 2019. Mentally healthy workplaces for fly-in fly-out (FIFO) workers in the resources and construction sectors – Code of practice: Department of Mines, Industry Regulation and Safety, Western Australia. Retrieved from: http://www.dmp.wa.gov.au/Documents/Safety/MSH_MHW_FIFO_COP.pdf.

Commission on Social Determinants of Health. 2008. Closing the gap in a generation: Health equity through action on the social determinants of health. Retrieved from: https://apps.who.int/iris/bitstream/handle/10665/43943/9789241563703_eng.pdf;jsessionid=EF4431FEEF17C4DD5CBD2A61CA23B4FE?sequence=1.

Connell R. 2012. Gender, health and theory: Conceptualizing the issue, in local and world perspective. *Social Science & Medicine* 74(11):1675–1683. doi:10.1016/j.socscimed.2011.06.006.

Global Business Coalition (GBC). 2013. Health & beyond tackling root causes. Retrieved from: http://archive.gbchealth.org/system/documents/category_74/468/Business-Action-on-Health-Awards_18.pdf?1367430418.

Global Reporting Initiative. 2018. GRI 403 occupational health & safety. Retrieved from: https://www.globalreporting.org/standards/gri-standards-download-center/gri-403-occupational-health-and-safety-2018/.

Hossain D, Gorman D, Chapelle B, et al. Impact of the mining industry on the mental health of landholders and rural communities in southwest Queensland. *Australasian Psychiatry: Bulletin of Royal Australian & New Zealand College of Psychiatrists* 21(1):32–37.

ICMM. 2008. Good practice guidance on HIV/AIDS, TB and malaria. Retrieved from: https://www.icmm.com/en-gb/publications/health-and-safety/good-practice-guidance-on-hiv-aids-tb-and-malaria.

ICMM 2009. Good practice guidance on health. *Impact Assessment*. Retrieved from: https://www.icmm.com/en-gb/publications/health-and-safety/good-practice-guidance-on-health-impact-assessment.

ICMM. 2009. Good practice guidance on occupational health risk assessment. Retrieved from: https://www.icmm.com/gpg-occupational-health.

ICMM. 2012. Overview of leading indicators for occupational health and safety in mining. Retrieved from: https://www.icmm.com/en-gb/publications/health-and-safety/overview-of-leading-indicators-for-occupational-health-and-safety-in-mining.

ICMM. 2013. Community health programs in the mining and metals industry. Retrieved from: https://www.icmm.com/website/publications/pdfs/social-and-economic-development/5788.pdf.

ICMM. 2015. Sustainable development framework: ICMM principles. Retrieved from: https://www.icmm.com/website/publications/pdfs/commitments/revised-2015_icmm-principles.pdf.

ICMM. 2018. Benchmarking 2018 safety data: Progress of ICMM members. Retrieved from: https://www.icmm.com/safety-data-2018.

IFC 2009. Introduction of health. *Impact Assessment*. Retrieved from: https://www.ifc.org/wps/wcm/connect/a0f1120048855a5a85dcd76a6515bb18/HealthImpact.pdf?MOD=AJPERES.

IFC. 2012. Guidance note 4 community health, safety and security. Retrieved from: https://www.ifc.org/wps/wcm/connect/dc3f4b80498007dca17ff3336b93d75f/Updated_GN4-2012.pdf?MOD=AJPERES.

IFC and EBRD. 2009. Workers' accommodation: Processes and standards A guidance note by IFC and the EBRD. Retrieved from: https://www.ifc.org/wps/wcm/connect/9839db00488557d1bdfcff6a6515bb18/workers_accomodation.pdf?MOD=AJPERES&CACHEID=9839db00488557d1bdfcff6a6515bb18.

IFC, Golder associates ltd and CARE Canada. 2004. HIV/AIDS guide for the mining sector. Retrieved from: https://www.ifc.org/wps/wcm/connect/topics_ext_content/ifc_external_corporate_site/sustainability-at-ifc/publications/publications_gpn_hivaids-mining.

ILO. 2014. Occupational safety and health in the mining (coal and other mining) sector. Retrieved from: https://www.ilo.org/global/topics/safety-and-health-at-work/industries-sectors/WCMS_219026/lang--en/index.htm.

Knoblauch AM, Divall JM, Owuor M, et al. 2018. Selected indicators and determinants of women's health in the vicinity of a copper mine development in northwestern Zambia. *BMC Women's Health* 18(1):62. doi: 10.1186/s12905-018-0547-7.

Lambe L. 4 June 2018. Ghana's unlikely marriage of mining and malaria control draws envious glances. *The Guardian*. Retrieved from: https://www.theguardian.com/global-development/2018/jun/04/ghana-unlikely-marriage-mining-malaria-control-envious-glances-obuasi.

Ling R, Kelly B, Considine R, et al. 2016. The economic impact of psychological distress in the Australian coal mining industry. *Journal of Occupational & Environmental Medicine* 58(5):e171–e176. doi:10.1097/JOM.0000000000000714.

Mawaw PM, Yav T, Mukuku O, et al. 2017. Prevalence of obesity, diabetes mellitus, hypertension and associated risk factors in a mining workforce, Democratic Republic of Congo. *Pan African Medical Journal* 28:282. doi:10.11604/pamj.2017.28.282.14361.

Mitjà O, Houinei W, Moses P, et al. 2015. Mass treatment with single-dose azithromycin for yaws. *New England Journal of Medicine* 372(8):703–710. doi:10.1056/NEJMoa1408586.

OECD. 2006. Measuring the effects of education on health and civic engagement. Retrieved from: https://www1.oecd.org/education/innovation-education/37425753.pdf.

Prüss-Ustün A, van Deventer E, Mudu P, et al. 2019. Environmental risks and non-communicable diseases. *BMJ* 364:l265. doi:10.1136/bmj.l265.

Rodriguez-Fernandez R, Farrand C, Cridland Raspail L, et al. 2016. Goliath Versus goliath: Leveraging private mining industry resources for public health gains in tobacco control. *World Congress of Cardiology Scientific Sessions*. doi:10.1016/j.gheart.2016.03.070.

Rodriguez-Fernandez R, Rahajeng E, Viliani F, et al. 2015. Non-communicable disease risk factor patterns among mining industry workers in Papua, Indonesia: Longitudinal findings from the Cardiovascular Outcomes in a Papuan Population and Estimation of Risk (COPPER) study. *Occupational & Environmental Medicine* 72(10):728–735. doi:10.1136/oemed-2014-102664.

Shandro AJ, Veiga M, Shoveller J, et al. 2011. Perspectives on community health issues and the mining boom–bust cycle. *Resources Policy* 36(2):178–186.

Stephens C, Ahern M. 2002. *Worker And Community Health Impacts Related to Mining Internationally: A Rapid Review of the Literature*. International Institute of Environment and Development and World Business Council on Sustainable Development, London. Retrieved from: https://pubs.iied.org/pdfs/G01051.pdf.

UNDP, IFC, IPIECA. 2017. Mapping the oil and gas industry to the SDGs: An atlas. Retrieved from: https://www.undp.org/content/undp/en/home/librarypage/poverty-reduction/mapping-the-oil-and-gas-industry-to-the-sdgs--an-atlas.html.

USAID 2012a. Audit checklist for an operating facility. *Emerging Infectious Diseases*. Retrieved from: https://2012-2017.usaid.gov/sites/default/files/documents/1864/Audit-Tool.pdf.

USAID 2012b. Proposed supplemental guidance to the IFC introduction on health. *Impact Assessment*. Retrieved from: https://www.usaid.gov/sites/default/files/documents/1864/Planning-Tool.pdf.

Viliani F, Edelstein M, Buckley E, et al. 2017. Mining and emerging infectious diseases: Results of the Infectious Disease Risk Assessment and Management (IDRAM) initiative pilot. *Extractive Industries & Society* 4(2):251–259. doi:10.1016/j.exis.2016.08.009.

WHO. 1946. Constitution. World Health Organization, Geneva. Retrieved from: www.who.int/governance/eb/who_constitution_en.pdf.

WHO. 2005. International health regulations. World Health Organisation, Geneva. Retrieved from: https://www.who.int/ihr/publications/9789241580496/en/.

WHO. 2018. Global status report on road safety 2018. World Health Organisation, Geneva. Retrieved from: https://www.who.int/violence_injury_prevention/road_safety_status/2018/en/.

WHO and UN Water. 2019. UN-water global analysis and assessment of sanitation and drinking-water (GLAAS). National systems to support drinking-water, sanitation and hygiene - global status report 2019. Retrieved from: https://www.who.int/water_sanitation_health/publications/glaas-report-2019/en.

Winkler MS, Krieger GR, Divall MJ et al. 2013. Untapped potential of health impact assessment. *Bulletin of the World Health Organization* 91(4):298–305.

GOAL 4

# Ensure Inclusive and Equitable Quality Education and Promote Lifelong Learning Opportunities for All

*Nicole M. Smith, Karin E. Olson Hoal, and John F.H. Thompson*

**CONTENTS**



## SUMMARY SDG 4: EDUCATION

| | |
|---|---|
| Key stakeholders | Community, youth, governments, workers, families |
| Direct mining contribution | Training, employment, skills development |
| Contribution to SDG 4 | Income to families, ability to afford and access children's education, transparency in operations, local engagement, future benefits to community, skills for other commercial growth, health benefits, security |
| Key regulations, policies | Major companies (e.g., members of ICMM) typically apply best practice wherever they operate globally; government regulations depend on country |
| Corporate policies, programs | Internal policy for trained workforce, commitments to stakeholder engagement, ICMM best practices |

## INTRODUCTION

Sustainable Development Goal 4 (SDG 4), "Ensure inclusive and equitable quality education and promote lifelong learning opportunities for all," includes wide-reaching goals for 2030 encompassing literacy and numeracy targets, access to safe, free, equitable, and quality education for all children and women, the elimination of gender disparities in education, and equal access to education and vocational training for persons with disabilities, indigenous peoples, and children in vulnerable situations (United Nations General Assembly, 2015). While few communities in any country enjoy the full benefits of education as outlined in SDG 4, the goals of providing equitable access to educational and life-long learning opportunities are critical for local and regional development.

Mineral resource development occurs where economic concentrations of minerals are found, whether in populated or remote regions within developed or developing countries. Traditionally, mining companies operating in remote areas built the towns and infrastructure, including schools, that were needed to provide for a workforce close to the site of mineral resource development—the mine. This approach is rarely used today, and modern mining operations more commonly utilize a fly-in-fly-out workforce in remote areas. However, companies do engage with local communities in a variety of ways. While many companies may have reduced their commitment to building mine towns and related infrastructure, the need to engage effectively with communities has increased dramatically in order to achieve high levels of social performance and social acceptance, as defined by the principals of Corporate Social Responsibility (CSR) and Social License to Operate (SLO) (e.g., United Nations Industrial Development Organization, 2019).

Increased engagement with communities has primarily included education in order to provide skills for employees in the operation, support community development, and in some cases, advance regional or national efforts. As we demonstrate in this chapter, many of these efforts go some way towards addressing SDG 4, although they are generally driven by the needs of the mining company. Nevertheless, changing societal expectations, as well as changes in the mining industry that include digital transformations, automation, and other technological efficiencies will require different approaches by mining companies in terms of community engagement and education. Therefore, we suggest that there are many opportunities to enhance educational activities related to mineral resource development in line with SDG 4 to the benefit of communities, the host country, and companies.

## BACKGROUND

Companies active in the mineral resource industries have an interest, along with the obligations of the governments of the countries in which they operate, in providing benefits to communities in exchange for the opportunity to extract resources. While these interests may not always be aligned, the company has a fiscal responsibility to its stakeholders to ensure the continuity of an operation, and therefore may view community benefits as a factor in revenue generation.

The International Council on Mining and Metals (ICMM) provides general guidelines to its member organizations for conducting baseline assessments of existing educational capabilities and for developing educational practices and skillsets (ICMM, 2019). These guidelines promote positive contributions such as investing in educational programs, carrying out recruiting and training initiatives, and collaborating with partners and development agencies to strengthen educational outcomes (ICMM, 2019).

Overall, there are two chief means by which resource companies participate in the education and life-long learning of communities. The first of these is indirect, through the allocation of rents and royalties paid to governments in accordance with international practice. The assumption is that governments will provide benefits, including education, to their citizens using the revenues generated from their mineral endowments (World Bank, 2009). While companies should advocate for human rights, the role of ensuring that governments provide education to their people is largely beyond the mandate of international resource companies. However, in Botswana, decades of consistent and reliable revenues from diamond mining activities have enabled the government to provide accessible and free education to children for their first ten years of schooling, with a focus on mathematics and science. The jointly owned entity of Debswana has ensured that both private industry (DeBeers) and government (Botswana) have roles in directing funds to community development. The country has been dedicated to training and upgrading the level of education for the next generation in anticipation of the post-mining economy (Republic of Botswana, 2015). In neighboring Namibia, petroleum and mineral companies support local universities in skill-focused training of Namibians for the energy and mineral sectors (Namibia University of Science and Technology Energy Institute and Centre for Oil and Gas, 2019). These initiatives contrast with other scenarios in resource-rich countries where monies may not be allocated to education (International Monetary Fund, 2011). In the Democratic Republic of the Congo, despite developments in primary school education, on average, only 18% of the students advance to secondary school (KPMG, 2014).

The second means by which companies engage in education and training is by establishing programs that directly benefit the communities in which they operate. There are numerous examples (e.g., University of Queensland, 2014; UNICEF, 2015; World Bank, 2014). In 2016, the World Economic Forum mapped the mining industries' activities to the Sustainable Development Goals (World Economic Forum, 2016). For SDG 4, they highlighted the Sierra Leone TVET program, a 2012 partnership between the German Development Corporation and London Mining, and the training, skills development, and scholarship program established by MMG at the Sepon mine in Laos (World Economic Forum, 2016). The global gem and jewelry industries also have contributed to community education efforts through instruments such as the Diamond Development Initiative. After many years of demand, the Gemological Institute of America published an educational guide specifically for artisanal and small-scale miners regarding practices and products that would provide more value from their efforts (Dirlam and Weldon, 2016). The Maendeleo Diamond Standards (MDS) provide artisanal and small-scale miners with guidelines for best practices and certification systems to bring their activities into the mainstream of the industry (Diamond Development Initiative, 2019). DeBeers has long held a policy of supporting local community education in every country in which they operate. The company focuses on building skills among women by supporting women-owned micro-businesses and a train-the-trainer approach in South Africa and Namibia through a partnership with U.N. Women and in providing 40 science-technology-engineering-math (STEM) educational scholarships to young women and girls in Canada (DeBeers, 2019).

Other examples have focused on local educational activities. In the early 1970s at Rössing Uranium on the west coast of Namibia, Rio Tinto broke from the traditional southern African mining practice of bringing male-only workers from the outside, and instead hired and trained local-area men and women. By using and training local human capital and investing in growing cottage industries related to farming and textiles, the company ensured a strong, educated, and economically diverse community around the mine at Arandis. Rössing received two World Aware Awards for Social Progress in 1994 for educational programs. The Crafts Development Program empowered local-area women in becoming self-sufficient and has since created a significant local industry, and the Maritime Training Centre provided courses in seamanship and developed a trained marine workforce for the country (Rio Tinto, 2016). In the 1990s, Rio Tinto similarly engaged native San youth in Namibia and Botswana in exploration activities, building skills, and providing income for the young people in the community (Hoal, 1996).

In Chile, a study by Martinez and Franks (2014) compared the impacts of social development and education practices at the Anglo American Los Bronces copper-molybdenum mine and the state-owned Codelco Andina copper mine. The study showed that education in particular was viewed by community stakeholders as an effective and positive benefit and long-term method of development. Anglo American's focus on education was therefore perceived more positively than Codelco's focus on other, shorter-term social capital activities (Martinez and Franks, 2014). In the U.S., Freeport McMoRan provides funding for K-12 STEM programs and grants to support teachers and schools in the rural communities of the American southwest in which they operate (Freeport McMoRan, 2019). Another approach to community development through education is to provide rural communities with the technical skills they will need to be hired in mining or other industries when companies enter their communities (e.g., Barefoot College, 2019).

Collectively, these examples demonstrate some ways in which minerals companies can support skill development among community members where they operate. Companies benefit from having a pool of skilled workers available, who may have been trained through industry programs or nongovernmental organizations, and who are prepared to join the company workforce. In short, the global and technical nature of the minerals industry requires a diverse and skilled labor pool that can be provided locally if workforce training is facilitated by industry, NGOs, and government. This can contribute to the overall aims of SDG 4.

## MINING'S REPORTED CONTRIBUTIONS TO SDG 4

Clear metrics for the contribution of the extractive industries to general education are difficult to determine. Quantitative measures of skill development, training, workshops, and community outreach may not be provided in annual reports, press releases, or NI 43-101 technical summaries. Instead, mineral resource companies typically report their contributions to sustainable development in their annual sustainability reports, which attract the attention of the general public. When sustainability information from minerals companies is made available to the public, there is often a lack of trust and credibility on behalf of the public that leads to misgivings about the nature of and drivers for company sustainability reports.

As part of a larger study on the sustainability reporting of mining companies in general, we evaluated the sustainability reports of eight companies over a six-year period (2012 to 2017). We initially selected the top 13 mining companies in terms of market capitalization, but for the analysis, we only used those with a formal sustainability report. We included eight companies in the final sample, which formed a representative sample of the primary metal industries (gold, copper, and iron). Our analysis quantified the metrics used by mining companies to disclose their contributions to SDG 4. We did not evaluate mining companies' performance.

In these reports, mining companies used the following five indicators to measure their contributions to education and life-long learning, key components of SDG 4 (Table 4.1):

(1) Programs for skills management and lifelong learning that support the continued employability of employees and assist them in managing careers;
(2) Average hours of training per year per employee by employee category;
(3) Percentage of employees receiving regular performance and career development reviews;
(4) Percentage of hours training (excluding health and safety) relative to the total hours worked (e.g., management, production, technical, administrative, cultural etc.);
(5) Number of employees per year that are financially sponsored by the company for further education.

Across the eight companies over a six-year period (for a total of 48 reports), the analysis revealed that 92% of the sustainability reports provided data for indicator 1: *Programs for skills management and lifelong learning that support the continued employability of employees and assist them in managing careers.* Following that, 74% of the reports provide data for indicator 2: *Average hours of training per year per employee by employee category,* with the types of training and the employee categories not specified. The focus on these two indicators suggest that mining companies placed a significant emphasis on educating their employees and developing their skills for jobs and careers within the mining operation, as well as beyond the life of the mine. Given the value generally placed on employment by local communities, as well as the ways in which employment opportunities fluctuate across the mining life

**TABLE 4.1**
**Number of Times (Percentage) Education-Related Indicators Appeared across the Sustainability Reports of Eight Mining Companies (48 Reports Total) in Each Year from 2012 to 2017 and over the Entire Six-Year Period**

| Indicator | 2012 | 2013 | 2014 | 2015 | 2016 | 2017 | Total |
|---|---|---|---|---|---|---|---|
| 1. Programs for skills management and lifelong learning that support the continued employability of employees and assist them in managing career endings | 100% | 88% | 88% | 88% | 88% | 100% | **92%** |
| 2. Average hours of training per year per employee by employee category | 75% | 75% | 75% | 75% | 75% | 71% | **74%** |
| 3. Percentage of employees receiving regular performance and career development reviews | 38% | 63% | 63% | 75% | 63% | 57% | **60%** |
| 4. Percentage of hours training (excl. health and safety) relative to the total hours worked (e.g., management, production, technical, administrative, cultural etc.) | 63% | 63% | 50% | 50% | 50% | 43% | **53%** |
| 5. Number of employees per year that are financially sponsored by the company for further education | 25% | 25% | 25% | 25% | 13% | 14% | **21%** |

cycle, companies should be responsible for developing skills and capabilities among their employees.

Over half (60%) of the company reports provided data on indicator 3: *The percentage of employees receiving regular performance and career development reviews*, and just over half (53%) provided data on indicator 4: *The percentage of hours training (excluding health and safety) relative to the total hours worked (e.g., management, production, technical, administrative, cultural etc.)*. Unfortunately, only 21% of the total reports over the six-year period provided data on indicator 5: *the number of employees per year that are financially sponsored by the company for further education*. This indicator is perhaps the most striking, clearly demonstrating the small proportion of employees who are typically sponsored for further educational opportunities.

These reporting trends suggest that companies are more focused on creating effective, safe, and healthy operations, preparing individuals to take on additional duties, and training their employees for mining and post-mining opportunities. While these data provide support for training and the life-long learning opportunities targeted in SDG 4, it appears that they largely serve the interests of the company and that there is room for improvement in terms of contributing to education and life-long learning. However, it is important to note that companies' education efforts beyond the mine site, such as contributions to and support for local schools and student scholarships, may be masked within other indicators that do not directly mention education. For example, companies commonly report on community projects in which they were involved; however, in their sustainability reports, there is generally little or no mention of the details of these community projects and whether or not education was a component.

## LESSONS LEARNED

Looking forward to 2030, the minerals industry has an important role to play with respect to education and SDG 4. There are many opportunities for companies to increase the degree to which they contribute to sustainable development by supporting educational and life-long learning opportunities in and around the areas in which they operate. While some companies have invested in schools and facilities as well as employee skill development and training, these efforts have largely appeared to be focused on inward-looking objectives to secure and maintain an effective workforce for the mining operation. In an era in which the availability of extractive industry workforce is dwindling, Deloitte (2018) and others have argued for an expansive view on what constitutes a minerals industry employee. Companies should not merely focus on developing an expatriate base for technical staff who have been traditionally trained. There are abundant new sources of human capital with which to engage for the relatively minor cost of skillset development. The economic benefits of skills training contribute to new commercial enterprises that help to grow a local economy and enable it to survive when natural resources are exhausted and operations cease. These efforts have positive impacts for families, communities, and local governments by maintaining employment, providing opportunities, and increasing capacity, hence supporting sustainable development.

Future avenues for education in developing regions of the world could include mineral resource companies providing online access and training opportunities, in addition to traditional secondary education. In many regions of the world, there simply is not enough capacity to build enough brick and mortar schools or institutions of higher education for the enormous need that exists. Education through distance learning in sub-Saharan Africa and India may be a better option for meeting that need where internet access exists and is growing, but if absent, traditional radio, television, and video training could still be supported by mining companies (Trines, 2018). Electrification and power generation required by resource companies for operations are among the benefits that the industry can and often does share with communities. These efforts can contribute to Sustainable Development Goal 9—*industry, innovation, and infrastructure*—by helping to build greater connectivity for the communities around mining operations. Resource companies that provide electricity and computers to schools and local government agencies can also help satisfy educational goals for more students. Online groups can be enlisted to aid companies in filling the gaps when they struggle to staff trainers. For example, the NGO 1MillionWomenToTech (1 Million Women To Tech, 2019) reaches thousands of women in developing countries from Africa to Asia, over 1000 in Nigeria alone, and resource companies that need workers trained with computer skills would find it useful to partner with such organizations rather than do the training themselves.

As the resource industries transform into high-tech, effective, smaller, and less wasteful operations, they have a real opportunity to support the communities in which they operate and to develop the technical and computational local workforce needed for the future—both in mining operations and in the communities that surround the operations. In so doing, the mining industry can play an important role in supporting the objectives of SDG 4.

## ACKNOWLEDGEMENTS

We would like to acknowledge and thank Dr. Sebnem Duzgun for her intellectual insights and Cansu Perdeli Demirkan and Aurora Waclawski for their assistance with the data analysis.

## REFERENCES

1 Million Women to Tech, 2019, https://1millionwomentotech.com.

Barefoot College, 2019, Modern Education for Rural Minds, https://www.barefootcollege.org/solution/education/.

DeBeers, 2019, Women's and Girls' Empowerment, https://www.debeersgroup.com/building-forever/our-impact/society/womens-and-girls-empowerment.

Deloitte, 2018, Future of Work in Mining: Attracting, Developing and Retaining Talent, https://www2.deloitte.com/content/dam/Deloitte/global/Documents/Energy-and-Resources/deloitte-norcat-future-work-in-mining.pdf, with Norcat, 10 pp.

Diamond Development Initiative, 2019, Maendeleo Diamond Standards Manual for Artisanal and Small-Scale Diamond Mining, April, 108 pp.

Dirlam, D.M. and Weldon, R., eds, 2016, *Selecting Gem Rough: A Guide for Artisanal Miners*, 1st ed., Gemological Institute of America, 53 pp. ISBN 0873110145, 9780873110143.

Freeport McMoRan, 2019, Schools as Partners, https://www.freeportinmycommunity.com/schools/.

Hoal, K., 1996, *Report of the Tsumkwe Diamond Exploration Project*, Rio Tinto Namibia Internal Report, Windhoek, Rio Tinto Namibia.

ICMM, 2019, International Council on Mining and Metals, 2019, SDG 4: Ensure Inclusive and Quality Education for all and Promote Lifelong Learning, http://www.icmm.com/en-gb/metals-and-minerals/making-a-positive-contribution/quality-education.

International Monetary Fund, 2011, *Beyond the Curse: Policies to Harness the Power of Natural Resources.* https://www.imf.org/en/Publications/Books/Issues/2016/12/31/Beyond-the-Curse-Policies-to-Harness-the-Power-of-Natural-Resources-24843.

KPMG, 2014, Global Mining Institute DRC Country Mining Guide.

Martinez, C., and Franks, D.M., 2014, Does Mining Company-Sponsored Community Development Influence Social License to Operate? Evidence from Private and State-Owned Companies in Chile. *Impact Assessment and Project Appraisal*,32(4), 294–303.

Namibia University of Science and Technology Energy Institute and Centre for Oil and Gas, 2019, http://nei.nust.na.

Republic of Botswana, 2015, Education and Training Sector Strategic Plan (ETSSP 2015-2020) April 2015.

Rio Tinto, 2016, Reflecting on 40 Years of Working for Namibia: The Next 10 Years 1986–1996, 17 pp.

Trines, S., 2018, Educating the Masses: The Rise of Online Education in Sub-Saharan Africa and South Asia. *World Education News and Reviews*, August.

UNICEF, 2015, Children's Rights and the Mining Sector. *UNICEF Extractive Pilot.*

United Nations General Assembly, 2015, Transforming Our World: The 2030 Agenda for Sustainable Development, 21 October 2015.

United Nations Industrial Development Organization, 2019, https://www.unido.org/our-focus/advancing-economic-competitiveness/competitive-trade-capacities-and-corporate-responsibility/corporate-social-responsibility-market-integration/what-csr.

University of Queensland, 2014, The Guide to Good Practices in Indigenous Employment, Training, and Enterprise Development, Sustainable Minerals Institute Centre for Social Responsibility in Mining.

World Bank, 2009, The Extractive Industries Value Chain, Extractive Industries for Development Series #3. *Africa Region Working Paper Series 125.*

World Bank, 2014, Human Capital for the Oil, Gas, and Minerals Industries.

World Economic Forum, 2016, Mapping Mining to the Sustainable Development Goals: An Atlas. http://www3.weforum.org/docs/IP/2016/IU/Mapping_Mining_SDGs_An_Atlas.pdf.

# GOAL 5 Achieve Gender Equality and Empower All Women and Girls

*Victoria R. Nalule*

**CONTENTS**



## INTRODUCTION

Achieving gender equality is a recognised goal at the national, regional and international level as stipulated in SDG 5.* From an economic perspective, gender equality entails access to equal opportunities and equal pay among men and women; socially and politically, it entails women having equal roles and responsibilities. Whereas some countries, especially in the developed world, are making good progress in the achievement of SDG 5 in all sectors, other countries, especially in the developing world, still have a long way to go. This discrepancy among countries in the achievement of SDG 5 is often attributed to the cultural, traditional and religious practices and beliefs that exist in different countries. Similarly, achieving gender equality globally in all economic sectors presents various issues. This is because some sectors, such as education and medicine, are more receptive to the inclusion of women, while other sectors, such as the mining sector, are predominately male-dominated. It is against this stark background that this chapter assesses the progress of SDG 5

* Internationally, there are various instruments including: The Convention on Elimination of All Forms of Discrimination Against Women (CEDAW); the Beijing Platform for Action; the United Nations Declaration on the Rights of Indigenous Peoples (UNDRIP); the Protocol on the Rights of Women in Africa (Maputo Protocol).

on gender equality in the mining sector. The main questions to be addressed in this chapter are:

- How are women affected by mining activities?
- Why is there a gender gap between men and women in the mining sector?
- What can be done to improve on the involvement of women in the mining sector?

In addressing the questions above, the chapter employs a three-step framework in the form of sections. The first section is the introduction, it analyzes the meaning of gender equality and how it impacts on other SDGs; the second section highlights how mining activities affect women specifically, analyzing the nexus between land access and mining; the third section provides the concluding remarks and recommendations to ensure the achievement of gender equality in the mining sector.

## What is Gender Equality and How Does It Impact Other SDGs?

Basically, mining and farming are some of the earliest areas of human enterprise and both men and women actively took part in these activities. However, compared to men, women do not have access to the same benefits that accrue from the mining sector, and yet they suffer the negative impacts associated with it. Before a discussion on the impact of mining on women, this subsection will highlight what gender equality means and how SDG 5 influences the achievement of other SDGs: in the same spirit we note how other SDGs impact on the equality gap between men and women.

### Understanding Gender Equality

People often mistake gender equality for sameness between men and women. This is not true, considering the visible physical differences and capabilities between the two sexes. Rather, gender equality is more concerned about men and women accessing the same rights, responsibilities and opportunities across all sectors of society. In this respect, therefore, we can measure gender equality by looking at the representation of men and women in different roles.*

Gender equality is a human rights issue and it is key in ensuring sustainable development and the achievement of all the other SDGs. This therefore explains why initiatives to achieve gender equality are now more than ever expressly supported in national legislation and international instruments. Equality of all humans, including male and female, is not only enshrined in international instruments, but also emphasized in national legislations. For instance, Article 21 of the Constitution of Uganda

* Measuring gender equality requires consistent statistics. In the EU for instance, Eurostat publishes gender statistics covering different sectors including education, health, labour market just to mention a few. See, https://ec.europa.eu/eurostat/statistics-explained/index.php/Gender_statistics. This can be accessed at http://hdr.undp.org/en/content/gender-inequality-index

is to the effect that all people are equal before the law irrespective of their gender. The article states that

> (1) All persons are equal before and under the law in all spheres of political, economic, social and cultural life and in every other respect and shall enjoy equal protection of the law. (2) Without prejudice to clause (1) of this article, a person shall not be discriminated against on the ground of sex, race, colour, ethnic origin, tribe, birth, creed or religion, social or economic standing, political opinion or disability.

This provision clearly emphasizes equality of all human beings in all spheres of life. In the mining sector, in order to ensure gender equality, emphasis must be put on ensuring that women have access to the necessary education, training and participation as men. They must not only be included in the negotiation of mining agreements but also in all the various stages of the mining sector.

International organizations have been a main driver for gender equality campaigns. Several programs have been set up to ensure adherence to gender equality requirements in the extractive industry. A case in point is the World Bank's Gender and Extractive Industries Program. The UN also takes gender equality very seriously and as such encourages member countries to include women in the oil and mining sectors.*

There are also various initiatives at a regional and international level. For instance, besides prioritizing gender equality in EU laws, institutions have been set up at regional level aimed at addressing gender issues. A case in point is the EU regional organization, where the European Institute for Gender Equality was established, and this is involved in various initiatives including establishing an online glossary of mainstream gender concepts and definitions.† Gender indexes are also used to measure gender equality in various sectors. For example, there are several gender indexes that exist, including the Social Watch Gender index;‡ the EU Gender Equality Index;§ the United Nations Gender Inequality Index.¶ Additionally, in 2016, the World Economic Forum introduced the Global Gender Gap Report aimed at highlighting gender-based disparities in different sectors.**

The UN through its trackers for SDGs, identified 9 targets and 14 indicators for SDG 5. Basically, the targets are aimed at identifying the goal, while the indicators represent metrics to track whether the targets are achieved or not.†† Some of the UN indicators are highlighted in Table 5.1.

---

* UNDP, UN calls for equal opportunities for women in the extractives sector in Mozambique (posted online on June 6, 2016) at http://www.mz.undp.org/content/mozambique/en/home/presscenter/pressreleases/2016/06/06/un-calls-for-equal-opportunities-for-women-in-the-extractives-sector-in-mozambique-.html

† See, https://eige.europa.eu/gender-mainstreaming/concepts-and-definitions. A similar glossary was compiled by the Council of Europe and this can be accessed at, www.coe.int/en/web/genderequality/publications

‡ This can be accessed at www.socialwatch.org/taxonomy/term/527

§ This can be accessed at https://eige.europa.eu/gender-equality-index/2015/IE

¶ This can be accessed at http://hdr.undp.org/en/content/gender-inequality-index

** The 2018 Global Gender Gap Report can be accessed at https://www.weforum.org/reports/the-global-gender-gap-report-2018

†† https://sdg-tracker.org/gender-equality

**TABLE 5.1**
**Selected UN Indicators**

| |
|---|
| Legal frameworks for gender equality and non-discrimination |
| Violence against women from an intimate partner |
| Time spent on unpaid domestic and care work |
| Women in political positions |
| Equal rights to land ownership |

We note that there is a lot to be done in order to achieve gender equality. For instance, in Table 5.1, with respect to the indicator on legal framework, there are several questions to be addressed such as:

- Does the law mandate equal pay for males and females?
- Does legislation explicitly criminalize marital rape?
- Do married men and women have equal rights to property and land?
- Does the law mandate paid or unpaid maternity leave?

The above questions are indeed necessary, but how do we incorporate these into the mining sector? Most mining legislation does not explicitly mention the need for equality among men and women in the sector. Additionally, we note that most mining communities are occupied by miners who in most cases lack advanced training or education. In this respect, therefore, we need to ensure that, besides the international and national legislation, the local and community laws should adequately address gender inequality in the mining sectors. These laws should also ensure that women have access to alternative agricultural and farming land in the event that they have to relocate due to mining activities.

## Definition of Key Terms

In understanding the impact of the mining sector on SDG 5, it is imperative to understand some key terms that influence SDG 5.

**Gender:** According to the *Cambridge English Dictionary*, this refers to the physical and/or social condition of being male or female.* Other scholars have introduced the concept of power which plays a major role when discussing gender in the mining sector. In this respect, gender has been defined as the culturally and historically developed concept that explains the power relationship between men and women.† This definition clearly brings out the role of cultural traditions in deciding what men and women can do. The aspect of having more power in society historically forced men to believe that women are less than them and this in effect shaped the socio-economic background of society. This therefore explains why the mining sector has historically been associated with strength and power – which is attributed to men.

* See, Cambridge Online Dictionary, https://dictionary.cambridge.org/dictionary/english/gender
† Connell, R.W., 1987. *Gender and Power.* Cambridge: Polity, pp. 183–185.

**Masculinity:** This is defined as the characteristics that are traditionally thought to be typical of or suitable for men.* Basically, the mining sector is often viewed as masculine and as such it is mainly dominated by men who traditionally believed that women and children are subordinate to them.

**Feminism:** This is defined as the belief that women should be allowed the same rights, power and opportunities as men and be treated equally.† It basically advocates for equality of all sexes in the social, economic and political sphere.‡ With respect to the mining sector, we note that women are the ones mostly affected by these activities and yet they do not get equal opportunities to benefit from the sector. The connection between mining and gender is well elaborated in the concept of ecofeminism, which looks at the connection between women and nature. The mining industry entirely depends on nature and yet we continue to experience the dominance of men in the extraction and production of different minerals, including gold, ore, diamonds, just to mention a few. This continues despite the introduction of technology in the extraction of these resources that could make it easy for women to contribute.

## How SDG 5 on Gender Equality Impacts/Is Impacted by Other SDGs

There are various SDGs that directly or indirectly have an impact on the achievement of Goal 5 on gender equality. These are briefly highlighted as follows:

a) **SDG 7 on energy access: gender, energy transitions, role of critical minerals**

Goal 7 of the UN Sustainable Development Goals (SDGs) emphasizes access to affordable, reliable, sustainable and modern energy for all.§ Basically, energy is essential in the economic development of a country as it is used in our everyday life for cooking, lighting, heating, transport, healthcare, just to mention a few.¶ We note that most of the challenges faced in the mining sector are in some way attributed to the lack of adequate modern energy; moreover there are now various initiatives to ensure utilization of renewables in mining.**

To clearly understand the connection between gender equality and energy access, we must have a clear understanding of what energy access entails. There is no globally agreed definition of energy access, although international organizations have been influential in developing this definition. In their 2017 report, the International Energy Agency (IEA) defined energy access as a household having reliable and affordable access to both

* See, Cambridge Online Dictionary, www.youtube.com/watch?v=V_Zr7IGBZNo

† See, Cambridge Online Dictionary, https://dictionary.cambridge.org/dictionary/english/feminism

‡ There are various types of feminism including liberal feminism, radical feminism, cultural feminism and eco-feminism.

§ See, 2015 UN Sustainable Development Goals, which can be accessed at https://sustainabledevelopment.un.org/sdg7

¶ Nalule, V.R., 2018. *Energy Poverty and Access Challenges in Sub-Saharan Africa: The Role of Regionalism*. Springer.

** Nalule, V.R., *Mining and the Law in Africa: Exploring the Social and Environmental Impacts*. London: Palgrave Macmillan.

clean cooking facilities and to electricity, which is enough to supply a basic bundle of energy services initially, and then an increasing level of electricity over time to reach the regional average.*

In simple terms, energy access means access to modern energy, which is reliable, sustainable and affordable. One may ask what modern energy entails? Modern energy can clearly be understood by distinguishing it from traditional energy and in so doing, we must look at the quality of energy used. For instance, regarding traditional energy, candles, kerosene and lamps are used for lighting; and firewood for cooking.† On the other hand, with respect to modern energy, electricity, natural gas and liquefied natural gas (LNG) are used for lighting and cooking, respectively.‡

The link between energy and gender equality is more visible in developing countries such as those in Asia and Africa. This is mainly due to the traditions and cultures in these countries that make women the major users of energy resources in local communities. Women are the ones most involved in social work such as cooking and other domestic work; this therefore implies that these women experience energy challenges differently and more severely from their male counterparts. Estimates indicate that women spend on average 1.4 hours a day collecting wood fuel and 4 hours for cooking.§ This is indeed one of the reasons why some scholars have argued that access to energy is gendered.¶ Ensuring access to modern energy therefore has the potential to improve the livelihoods of women who depend on these energy resources for their day-to-day activities in developing countries, and as such contributes to achieving the SDG target of gender equality.

Having highlighted the influence of SDG 7 relating to energy access on gender equality, it is important to point out the nexus between energy and mining. The global move towards a low carbon economy and tackling climate change as emphasized in the 2015 Paris Agreement, has necessitated more investments in renewable energy. Metals such as cobalt, lithium and rare earths are key in the expansion of renewable energy systems. These metals are now in very high demand and yet they are concentrated in a small number of countries. This presents an opportunity for women involved in the mining of these critical minerals. At the same time, this is likely to expand gender inequality, especially in countries that haven't addressed this issue at the national level. Additionally, mining of these critical minerals not only increases environmental challenges but also conflict

* See International Energy Agency: World Energy Outlook, Special Report, 2017, p. 21. It can be accessed at https://www.iea.org/publications/freepublications/publication/WEO2017SpecialReport_EnergyAccessOutlook.pdf, last accessed on 12 November 2019.

† Nalule, V.R., 2020. Transitioning to a Low Carbon Economy: Is Africa Ready to Bid Farewell to Fossil Fuels? In *The Palgrave Handbook of Managing Fossil Fuels and Energy Transitions* (pp. 261–286). London: Palgrave Macmillan.

‡ Nalule, V.R., 2019. Energy Access in Sub-Saharan Africa: In *Energy Poverty and Access Challenges in Sub-Saharan Africa* (pp. 21–39). London: Palgrave Macmillan.

§ Nalule, V.R., 2019. Energy Access in Sub-Saharan Africa. In *Energy Poverty and Access Challenges in Sub-Saharan Africa* (pp. 21–39). Cham: Palgrave Macmillan.

¶ Ibid.

with other land use. As will be discussed in the next section, women are mostly dependent on the land and as such are heavily affected by mining activities on their land.

b) **Link between SDG 5 and other SDGs**
Other SDGs, including Goal 4 on quality education, are also linked to gender equality. Unlike in developed countries where girls and boys have equal opportunities to go to school, in some developing countries girls are often discouraged from taking up higher education mainly due to cultural and traditional beliefs. Some tribes and cultures require girls to get married at an early age and they therefore never get a chance to acquire the education they need. Additionally, due to their social role, girls miss out on school because they must stay home to help with family chores including cooking and farming, just to mention a few. The link between gender and education therefore lies in the fact that we cannot achieve quality education for all if women are not given a chance to attain this education. There are indeed more links between SDG 5 and the other SDGs including 1 on poverty eradication; 2 on hunger eradication; 3 on good health, just to mention a few.

There are various issues to discuss with respect to gender equality; this chapter, however, focuses on gender equality from the perspective of the mining sector. This will be analyzed in the next section.

## GENDER AND MINING: WHAT IS THE PROBLEM?

### Nexus between Gender and Mining

The mining sector is increasingly becoming very important considering the escalating demand for critical minerals that are used in renewable energy systems globally. Basically, mining refers to the act of removing substances such as coal or metal from the ground.* In this short paper, gender inequality in the mining sector will be discussed drawing examples from African countries.

The mining sector is essential in the economic development of many resource-rich African countries as it has the potential to finance infrastructural developments that are much needed in Africa.† This is due to the fact that various minerals are used as raw materials in different sectors including construction and transport. Most importantly, mineral materials also play an important role in industrialization and urbanization, which are now rapidly escalating on the continent. Africa is home to 30% of mineral reserves and these significantly contribute to exports and tax revenues for most countries on the continent.‡

In understanding the extent of gender inequality in the mining sector, we must note that mining is carried out on both a large-scale and a small-scale. In this respect

* See, Cambridge Online Dictionary, https://dictionary.cambridge.org/dictionary/english/mining. Last accessed on January 13, 2020.

† Nalule, V.R., 2020. *Mining and the Law in Africa: Exploring the social and environmental impacts.* London: Palgrave Macmillan.

‡ Nalule, V.R., 2020. *Introduction to Mining in Africa. In Mining and the Law in Africa* (pp. 1–17). London: Palgrave Macmillan.

therefore, the challenges women face depend on the type of mining involved. While large-scale mining (LSM) involves companies, who employ advanced technology and enormous capital to extract minerals, artisanal mining or small-scale mining (ASM) on the other hand involves the use of rudimentary methods by individuals who are not officially employed by a mining company.*

Unlike LSM, ASM attracts a significant number of women in the sector. In Table 5.2, we highlight the challenges women face in ASM.

As illustrated in Table 5.2, the challenges women face in ASM can be categorized into economic challenges, social challenges, and political challenges.

Economically, women have no access to finance, and this limits them from acquiring the land, machinery and equipment needed in ASM activities. Additionally, lack of finances has forced many women to work without the necessary protective gear and this has exposed them to dangerous diseases. A case in point is salt mining and gold mining. I had a chance to visit both the ASM gold site and the ASM salt site in Uganda. For gold mining, the use of mercury was evident from all the miners, especially women.† One striking feature is how most women involved in ASM do not have protective equipment such as gloves and boots. For instance, in the salt mines I visited in Katwe, Uganda, most women involved in salt mining did not have boots or gloves. They explained that lack of such protective equipment was associated with various health impacts, including being exposed to ammonium gas, which negatively impacted their private parts, exposing them to various dangerous diseases.‡

Socially, the cultural and traditional expectations of women hinder their active participation in the mining sector. For instance, in some cultures in Africa, women are not allowed to get involved in certain stages of mining including digging of ores or sale of minerals. In some cultures, it is believed that it is bad luck for a woman to be involved in the first and last stage of mining. Additionally, such activities have often been labeled as "men's work". This therefore undermines the role of women in ASM even though women play a significant role in the sector. Even though not many women are seen digging pits, they are often involved in other essential activities in mining communities including catering, laundry, cleaning and many other goods and service provision

**TABLE 5.2**
**Challenges Women Face in ASM**

| |
|---|
| Economic challenges such as lack of finance to invest in ASM projects; protective gear; modern equipment and machinery |
| Social challenges such as traditional and cultural expectations which in most cases deny women a right to fully participate |
| Political challenges such as being victims of wars in mineral-rich countries |

* Hilson, G., Goumandakoye, H., and Diallo, P., 2019. Formalizing Artisanal Mining 'Spaces' in Rural Sub-Saharan Africa: The Case of Niger. *Land Use Policy*, 80, pp. 259–268.

† See video I recorded from the gold mining site in Uganda, www.youtube.com/watch?v=HuWmudeSz_A

‡ For an overview of the impact of salt mining on women in Uganda, watch the video I recorded from the field, www.youtube.com/watch?v=RKFP6DswMrM&t=491s

around ASM. Women are also involved in other mining stages including processing and delivery, and some do get involved in the exploration and mining of the metals. We note that mining practices have for long been an area of gendered work and women have for long been involved. The role of women in the mining sector is well documented; for instance two women in Ghana are celebrated for their influence in the country's sector before colonization and the introduction of machinery in the mining sector.*

The above notwithstanding, there is a need to advocate for strong participation of women in the whole cycle of mining and related activities, from exploration to consumption. We also note the domestic roles of women including cooking, raising children, looking after the household, just to mention a few. These domestic roles not only make it hard for women to be actively involved in ASM but have also made it hard for some women to acquire the necessary education and training that are essential in taking up some formal jobs in the mining sector.

Politically, women have often been victims of rape in war-torn resource-rich areas where rape is sometimes used as a weapon of war. The mining sector is generally characterized by gender-based violence, especially in conflict-intense countries such as the DRC. Rape is often used as a tool to oppress the local communities. Most people see gender in a natural sense; hence they naturally view men as masculine and women as feminine.† This has escalated gender-based violence, as women are viewed as a weaker sex and an easy target for political violence. Additionally, the political unwillingness of some countries to amend all the relevant laws to favor women also makes it impossible for women to benefit in the mining sector.

There are indeed various issues that have barred women from gaining from the mining sector. It is imperative to note that, despite the barriers outlined earlier, women are involved in all the processes of mining including production and marketing; however these roles remain invisible. For instance, traditionally, as evidenced in India, women worked closely with men in the mining sectors, such as coal, just as they did in the plantations.‡ Currently, women's contribution in and around the mines is visible on site and it demolishes the myth that the mining sector is male-dominated. The traditional reference to mining as a male-dominated sector is understandable because mining is basically risky and hazardous and as such most people think it is suited for men. It is mostly men who go down the mines, risking their lives to feed their families; this has therefore created a strong solidarity among miners which traditionally associates manual labour with masculinity. The isolation, risks and solidarity of earlier mines have endowed manual labour with the attributes of masculinity.§ In this respect, there are various issues that policymakers must continue addressing if women are to equally benefit from the mining sector. This article will mostly focus on the issue of land access and this is discussed in the next subsection.

---

* Akurang-Parry K.O., 2006. Making a Difference in Colonial Interventionism in Gold Mining in Wassa Fiase, Gold Coast (Ghana): The Activism of Two Women, 1874–1893. In Gier J.J., Mercier L. (eds) *Mining Women*. New York: Palgrave Macmillan.

† See, Macdonald, C. 'The Role of Gender in the Extractives Industries', WIDER Working Paper 2017/52. Helsinki: UNU-WIDER (2017). Accessed from: www.wider.unu.edu/publication/role-gender-extractives-industries on March 12, 2019.

‡ Lahiri-Dutt, K., 2007. Roles and Status of Women in Extractive Industries in India: Making a Place for a Gender-Sensitive Mining Development. *Social Change*, 37(4), pp. 37–64.

§ Lahiri-Dutt, K., 2007. Roles and Status of Women in Extractive Industries in India: Making a Place for a Gender-Sensitive Mining Development. *Social Change*, 37(4), pp. 37–64.

## Land Access: How Does It Escalate Gender Inequality in the Mining Sector?

Access to land is essential for the successful operation of oil, gas, energy and mining projects. However, there are often social, environmental and economic issues associated with acquiring land for these projects. Socially, many people are relocated to unfavorable places; economically, local communities are not given adequate compensation; environmentally, extractive industries are associated with pollution, which negatively impacts on the agricultural and fishing industries relied on by the local communities, especially women.* This subsection highlights how poor land management escalates gender inequality in the mining sector.

Basically, ownership of land is key in determining who owns the legal rights and title of such land. In some countries, citizens have a right to own land, while in other jurisdictions, the ownership of land is vested in the State.† In Uganda, for instance, the ultimate ownership of land is vested in the country's citizens. In this respect, the supreme law of the land, which is the Constitution under Article 237(1) and (3), vests the ownership of land in the citizens of Uganda. The Article further outlines the different tenure systems under which the land can be held.‡

Due to cultural, social and economic challenges, women are in most cases not in positions to own land. Land is key in ASM and LSM activities. As discussed in the previous subsection, women often contribute in ASM, but they do not benefit as much as men. The situation is different for the few women who own the land where the ASM projects take place, as these do get rent and royalties from the miners.

With respect to LSM, we note that the landowner only gains from the compensation that he/she receives. The owner of the land is entitled to compensation if oil, gas and mining activities are to be carried out on his/her land. The reasoning behind compensation is based on the fact that a mining title does not per se extinguish the rights of the landowner, rather it grants the owner of a mining licence a right to exploit the natural resources present in the subsoil, which definitely requires access to the surface.§ In Uganda for instance, natural resources, including hydrocarbons and minerals, are held by the government on behalf of the people. In this respect, the State will intervene to ensure that the landowners are fully compensated. If no agreement can be reached then the State may move to expropriate the land for mining purposes, in which case the landowner will still be entitled to compensation.

Connecting the issue of land to gender equality, we note that in most cases it is men who are owners of the land in question. In this respect, men highly benefit from LSM, as they get all the compensation that comes with it. Additionally, the married women who are entitled to gain from the compensation on their marital home are

* Nalule, V.R., 2020. Social and Environmental Impacts of Mining. In *Mining and the Law in Africa* (pp. 51–81). London: Palgrave Macmillan.

† Nalule, V.R., 2020. Introduction to Mining in Africa. In *Mining and the Law in Africa* (pp. 1–17). London: Palgrave Macmillan.

‡ The Constitution of Uganda, 1995, as amended.

§ Nalule, V.R., *Mining and the Law in Africa: Exploring the Social and Environmental Impacts*. London: Palgrave Macmillan.

often left out, as all the compensation goes to the husband. This is an issue that must be addressed in both mining legislation and land legislation.

Gender inequality, especially with respect to land access and extractives, has been documented in countries such as Kenya, where the cultural traditions of the affected people such as the Turkana, make it hard for women to benefit from the extractive industries.* Customary land ownership in African countries such as Uganda and Kenya also makes it hard for women to benefit from compensation. The land laws, however, do protect family land. In this respect, a spouse is required to give her written consent before the family land is mortgaged or sold.†

We should also be aware that most land where mining is carried out is in rural areas with communities, especially women, who depend on agriculture, farming and fishing. If these women are not adequately compensated, then their livelihood is automatically negatively impacted. Additionally, mining activities have often led to relocations; in this respect, therefore, it is essential to ensure that relocation plans are put in place that favour women and children.

## CONCLUSION: WAY FORWARD TO ADDRESS GENDER INEQUALITY IN THE MINING SECTOR

There are various issues associated with gender and mining as highlighted in the previous section. Several initiatives on the local, national, regional and international level have been taken in different countries to generally address gender inequality in all sectors. In the mining sector, the issue of gender has been put at the forefront. Nevertheless, not much progress has been seen to practically ensure that women benefit equally with men from the mining sector. In this respect, therefore, there is a need for more practical solutions as highlighted in Table 5.3.

**TABLE 5.3**
**Practical Solutions for Gender Equality in the Mining Sector**

| |
|---|
| Sensitizing mining communities, organizations, government ministries to the need for equal opportunities in the mining sector |
| It is essential to understand the social, historical and traditional influences that are likely to have an impact on gender equality in the mining sector |
| There is need for more debate on how to further include women in the mining production process |
| Involve women in the negotiation of mining projects at the local level |
| Amend all the relevant laws, including land laws, to ensure that they put women at the forefront |
| Focus on practical initiatives that actively involve women in the mining communities |

* Mutua, E, 'Report on the Policy, Legal and Regulatory Framework in The Extractive Industry of Kenya: Consideration of the Interests of Local Communities, Women, Youth and The Vulnerable, and The Environmental Sustainability of The Industry' April 2014. UNDP.

† Land Act 1998 Cap 227 Laws of Uganda. Uganda Land Act 1998. Available at http://www.ulii.org/ug/legislation/consolidated-act/227

## BIBLIOGRAPHY

Cameron, P.D., and Stanley, M.C., 2017. *Oil, Gas, and Mining: A Sourcebook for Understanding the Extractive Industries*. The World Bank.

Lahiri-Dutt, K., 2007. Roles and Status of Women in Extractive Industries in India: Making a Place for a Gender-Sensitive Mining Development. *Social Change*, 37(4), pp. 37–64.

Macdonald, C., 2017. The Role of Gender in the Extractives Industries. *WIDER Working, Paper 2017/52*. Helsinki: UNU-WIDER. Accessed from: https://www.wider.unu.edu/publication/role-gender-extractives-industries on 12th March 2019.

Nalule, V.R., 2018. *Energy Poverty and Access Challenges in Sub-Saharan Africa: The Role of Regionalism*. Springer.

Nalule, V.R., 2020. *Mining and the Law in Africa: Exploring the Social and Environmental Impacts*. Springer Nature.

Hilson, G., Goumandakoye, H., and Diallo, P., 2019. Formalizing Artisanal Mining 'Spaces' in Rural Sub-Saharan Africa: The Case of Niger. *Land Use Policy*, 80, pp. 259–268.

GOAL 6

# Ensure Availability and Sustainable Management of Water and Sanitation for All

*Andrea Juárez*

## CONTENTS



## ANALYSIS

The sixth sustainable development goal must be seen as part of the group of goals related to the social dimension of the SDGs. This specific SDG aims to "ensure access to safe water sources and sanitation for all" since it is a human right that billions of people are challenged daily to have access to. And it is the goal that will be examined throughout this chapter.

Central America is the delightful tropical weather region that is "bordered by Mexico to the north, Colombia to the southeast, the Caribbean Sea to the east and the Pacific Ocean to the west and south. It consists of seven countries; Belize, Costa Rica, El Salvador, Guatemala, Honduras, Nicaragua and Panama" (The Free Encyclopedia, n.d.), which together have an estimated population of 47.4 million people* within an area of 523,780 km;[2] where at least, 17.54% of the population have no access to basic sanitation services and 5.4% have no access to drinking water services. Of these seven countries, Guatemala is the most populated, with 17.2 million people as of 2018 (see Figure 6.1); it has high water availability (93,388.5 million $m^3$ of fresh water annually) (Ianrna-URL, 2012) and is also the country with the worst basic sanitation services. Its national government invested 0.8% of the GTQ 987.5

* As of 2018.

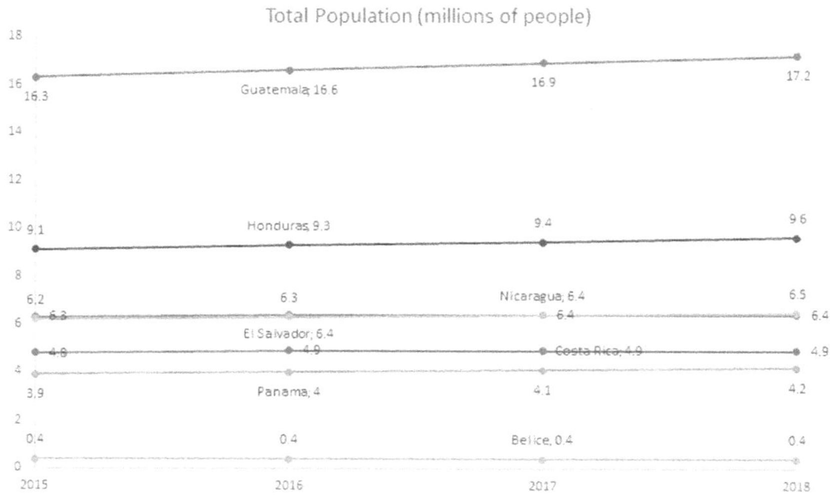


**FIGURE 6.1** Total population (millions of people) by country. Source: own creation with data extracted from the World Bank.

million (about US$ 957.2 million) budget, specifically and exclusively to the seven national development priorities, in water and natural resources.

Why is water scarcity and inadequate sanitation a problem for the development of the people?

The United Nations explains that "water scarcity, poor water quality and inadequate sanitation negatively impact food security, livelihood choices and educational opportunities for poor families across the world" (United Nations, United Nations Development Programme, n.d.).

According to the UNWATER program (United Nations, UNWATER, n.d.) many communities around the globe rely on water sources that are far away from their homes and, normally, it is the socially assigned job of women and young girls to gather water daily, leaving them vulnerable and exposed to different dangers. In Guatemala, because of the large distances between houses in the rural areas, it is very difficult for the municipalities to provided water services efficiently. Only 60%* of Guatemalan houses have access to water services and, in some departments, it is only 22.6%.

This situation forces many women to look for an alternative natural source of water, which, most of the time, is not a reliable source, so they are also exposed to different types of diseases. "Each day, nearly 1,000 children die due to preventable water and sanitation-related diarrheal diseases" (United Nations, Clean Water and Sanitation, n.d.). In Guatemala, the mortality rate by diarrhoea has been decreasing in recent years but incidents are still occurring in the city and are highest in rural areas (see Figure 6.2).

This situation has a negative externality that affects directly the educational opportunities of both boys and girls; historically is the women's job to take care of

* By 2011.

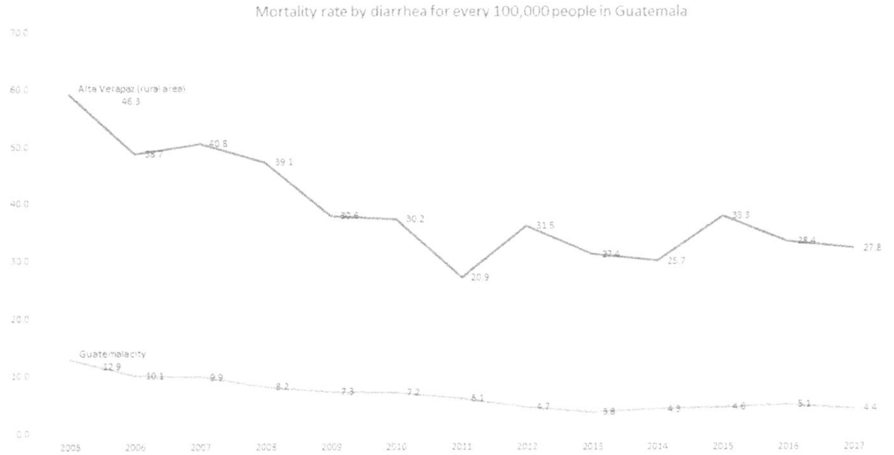


**FIGURE 6.2** Mortality rate by diarrhea for every 100,000 people in Guatemala. Source: own creation with data extracted from the INE (ine.org.gt).

the household, so girls have to invest a lot of time, mostly school time, in fetching water for the family and, because of the poor quality of the water, the children suffer constantly from contaminated water-related diseases, and this keeps them from being able to go to school.

One source of the contamination in the water is due to a lack of sanitation in many communities, and because of this people are forced to use communal latrines* or defecate out in the open, which only makes the risk higher of faecal matter contaminating the nearby food and water sources. "2.4 billion people lack access to basic sanitation services, and 892 million people practiced open defecation" (United Nations, Clean Water and Sanitation, n.d.). By 2011 only 42.4% of homes in Guatemala had a toilet connected to a drainage system, 36.2% made use of a latrine and 7.8% of the homes practiced open defecation. (Guatemala, n.d.) Added to this is the fact that many communities have little or no hygiene behaviour, and when they do, they do not always have access to the hygiene-related products they need.

This is the story of many people around the globe, especially people in developing countries, including the seven in Central America (see Figure 6.3).

"Lower-income countries have the widest gap to bridge in meeting SDG targets" (Onil Banerjee, 2019), starting with the fact that a lot of monetary resources are needed to meet the goals. The global monetary estimation of the resources that will be needed to target the 17 goals is from US$13–14 billion annually (Unidas, 2017).

Central American countries do not have the immediate liquidity to invest huge amounts of money in infrastructure and all the things needed to meet all the goals at the same time, so the advice that has been given to them and that governments are putting into practice is: "Lower-income countries need to prioritize which SDGs to focus on in alignment with national development priorities" (Onil Banerjee, 2019).

* "A simple toilet such as a hole in the ground" (Cambridge Dictionary, n.d.)

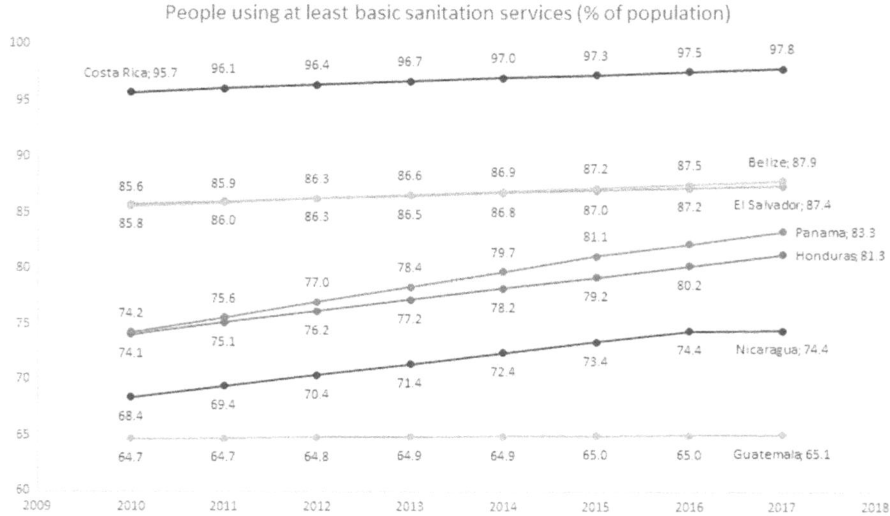


**FIGURE 6.3** Percent of population using sanitation services. Source: own creation from data extracted from the World Bank: World Development Indicators.

Another option is what Daniel Titelman stated, about the Latin American and Caribbean countries, in a sustainable development forum (Unidas, 2017). He said that the first step to implementing the goals is to gather the resources. He explains that there are four sources:

- Public: increase taxes and improve the structure of taxation to maximize income
- Private: resources perceived by exports and remittance
- Domestic: enforce laws and institutions to diminish tax evasion, and
- External: direct the donations and aids better to the areas prioritized

Titelman has a point, but his ideas and solutions will take time.

Central American governments have another option to take into account in order to meet the 2030 goals; they can turn to metal mining companies to form a synergy.

This sounds easier than it will be. But, the different synergy opportunities with other institutions can make it possible and successful.

But, let's take it one step at a time.

## How Could This Work? The Case for New Mining Projects

If a group of people is going to play a game, the rules of the game must be clear, must apply to everyone equally and they must be set beforehand. They cannot be changed along the way, and every player must agree to the rules if they want to play.

Guatemala has a weakness in the water-related legal framework; basically the norms that indicate who is the owner and who is responsible for the resource according to the ownership of the land, how much of it can be used at each time of the year and how to make responsible use and treatment of the resource, lack of clarity and lack of supervision. These weaknesses in the norms make the government's and the communities' roles and obligations, of providing reliable and near-water sources for the people, almost impossible. These weaknesses are also a big obstacle for the development of the responsible private sector too, and a window for the irresponsible water usage and disposal coming from both communities and the private sector. So, the first step would be to improve these public policies.

In this first step, the government could seek a first synergy with both private and public universities. With their help they could design a plan to improve and clarify, for both public and private sector, the formal water-related institutions currently in existence. The main focus point would be to develop a plan that would allow the efficient use, storage, conservation, distribution and treatment or recycling of water to benefit economic, environmental and social development, based on the history, needs and structure of the country. In other words, find a sustainable solution for the use, distribution and treatment of water in the different parts of the country.

Once this water-related legal framework is clear and established, another strategy and synergy can arise. In the mining regulation laws of Guatemala, the rules of the game are established. If, for example, a mining company wants to enter the country, they need to fulfill the requirements established in the mining law (Guatemala G. d., n.d.). A new article could be negotiated and here the new synergy would take place – a synergy between the Guatemalan government, the interested mining companies, the universities and maybe even civil society – to create this new article that would encourage all parties to cooperate.

The idea of this negotiation and synergy is to make a new article or game rule that is fair, clear and attractive to everybody; convenient and beneficial for local government and the community, because not only would they receive royalties, taxes and jobs from the foreign investment but also, the foreign company would help the country meet the sixth SDG, improving the quality of life of the people. It would be an article that significantly reduced the uncertainty of expectant foreign investors at a moderated cost, increasing the interest of big companies waiting to enter the country; an article that would also respect the community's interests and informal institutions and, last, it would be a negotiation in which the universities' representatives would play a mediation role to assure this break-even point.

The new game rule created by this synergy would then be added to the existing mining law, legislation that would require the incoming company to take a leading role in a water infrastructure development project. It would be supervised and supported by the communities that surround the project area and who would directly benefit from it, and by a government Institution, like the Ministry of Energy and Mining (MEM, according to the Spanish name), which would ensure that the liabilities were met. In this way, the companies would take into account, from the initial project planning, that they have to include the design, construction and execution of a shared, efficient and sustainable water infrastructure that will cover both the project and the neighbouring communities, and the local government and community would have another incentive to reduce the uncertainty for foreign investment companies interested in entering the country, resulting in a zero–sum game.

## What About the Already Existing Mining Projects?

The now existing and already operating metal mining companies in Central America are located in remote rural areas, where investment in water and sanitation infrastructure is urgently needed.

Following the example of Guatemala, the best-known active exploration and in-construction phase mining projects in the country are;

- El Escobal project from Pan American Silver in Santa Rosa, Jalapa and Jutiapa, and
- Cerro Blanco project from Entre Mares de Guatemala in Jutiapa.

Table 6.1 shows the percentage of homes with water access by 2011 in the communities where the mining projects are present.

Table 6.2 shows the percentage of homes with the different sanitation services available by 2011 in the communities where the mining projects are present.

The communities where the projects are present are in urgent need of water and sanitation infrastructure improvement, as has been shown. Both these foreign projects are having trouble initiating their operations because of community

**TABLE 6.1**
**Water Access in Communities with Mining Presence**

| Community | % of homes with water access by 2011 |
|---|---|
| Santa Rosa | 58.8 |
| Jalapa | 54.1 |
| Jutiapa | 51.5 |

*Source:* own creation with data extracted from the INE (ine.gob.gt).

**TABLE 6.2**
**Percent of Homes with Sanitation Services in Communities with Mining Presence**

| Indicators/Community | Santa Rosa | Jalapa | Jutiapa |
|---|---|---|---|
| Toilet connected to drainage system | 20.6 | 24.9 | 27.5 |
| Toilet connected to septic tank | 19.9 | 9.6 | 7.2 |
| Washable lavatory | 9.6 | 6.8 | 16.3 |
| Latrine | 38.9 | 38.7 | 21.1 |
| Has none | 10.9 | 20 | 27.9 |

*Source:* own creation with data extracted from the INE (ine.gob.gt).

and government disturbances. With these already existing projects, an agreement could be made to align the objectives of the mining companies with the objectives and need of the communities: infrastructure in exchange for operations permit.

## Is It Possible to Align the Objectives of the Mining Companies and the Communities?

Taking a look at the mining companies' strategies and goals, it is very easy to find that one of the priorities is to generate value and to have a competitive advantage by being an environmental, social and a governance leader (Newmont, n.d.). Also, corporate strategies aim to contribute to the United Nations SDGs in the countries they find themselves in:

> Pan American Silver is committed to the principles of sustainable development and the values essential to advancing the SDGs and the 2030 Agenda. We aim to conserve and recycle water, support community water and sanitation infrastructure, and are helping to build community-based capacity in this area. We continue to look for opportunities to increase the efficiency of our wastewater treatment systems.
>
> **(Silver, 2018)**

This shows a not-so-evident but existent connection and alignment in both parties' objectives. To meet their corporate goals – supporting the SDGs – the foreign investment companies are forced to direct their sight and effort in improving their neighbouring community's situation. Because if the SDGs are met, the communities' quality of life improves and the company gains in value. And in order to meet the SDGs and improve their quality of life, the communities and local government need the project to continue and to invest in infrastructure. So, the answer to the question is that it is not only possible to align the objectives of the mining companies and the communities but, the alignment between the corporate strategies and the communities already exists; just that, maybe, the people have not given it deep thought. If both parties were to realize this, and negotiate it correctly, it could be an incredible opportunity for countries like Guatemala to receive the help needed to meet the goals and, at the same time, for mining companies to expand to other areas of different central American countries.

This is an opportunity to create a synergy between the different stakeholders: the mining companies, government at national and local level and civil society groups. This synergy would consist of teamwork between the mentioned entities, aiming at the improvement of safe water availability and sanitation problems that the rural neighbours of the mines could be facing, in exchange for stability and reduction (if not elimination) of work disturbances.

The main idea would be to further align private goals and community needs, and combine knowledge and monetary and human resources, to design a plan and to successfully develop new projects, contributing in this more efficient way to the community's development and education to move the country closer to meeting the sustainable development goals.

## How Could This Work or Be Managed If the Game Has Already Started?

A perfect example is the Marlin mine in San Marcos, Guatemala, owned by the Newmont Goldcorp company. As part of the commitment made to the communities before the construction started, they invested in improving the infrastructure of the schools, libraries and roads. They also invested in building a new sports centre and a fully equipped health facility. With these projects, the mining company had already fulfilled most of the promises they made to help the neighbouring communities prosper (the remaining commitments would be met when the mine started the production phase and provided job for many community members). During the construction phase the company invested millions of dollars more in infrastructure for electricity, roads, streets, bridges and common rooms as part of their goal to improve access to the mines' neighbouring communities. The communities benefited from this also as a positive externality. The mine then entered the production phase from 2005 to 2017. But, in 2016 something happened and an opportunity to help the communities further arose. The Guatemalan government had formed a National Urban and Rural Development Board in 2014, called the national development plan: K'atun. The board was created to organize and manage development programs between public and private entities. In 2016 this program, together with the Inter-American Commission on Human Rights (CIDH in Spanish), demanded that a neighbouring community of the mine perform 18 water-related projects, directed specifically to improving the clean water and sanitation difficulties that the people where facing. With these projects in hand, the community turned to the Marlin mine and asked for their help to fulfill the projects. The Marlin mine, in spite of having already accomplished most of their planned projects for the communities, decided that they were going to help in some of the projects. Together they came up with a plan to work side by side, supervising each other, in 11 projects, in order to improve the situation of the communities as a team. This plan consisted mainly in a division of interlaced tasks. The mine bought the materials necessary for the projects, the community brought the workforce that built the necessary infrastructure, the mine supervised the whole process and, once it was finished, the community and the mine presented the results to the Guatemalan government (projects are still being reviewed to this day). The mining company invested an extra of GTQ 102.0 million (US$ 13 million) for these projects (Newmont-Goldcorp, n.d.). Thanks to this synergy between the private and public entities, many families got access to piped water and their living conditions improved significantly, and Newmont-Goldcorp's image improved as a result of this action; there is no such thing as a free lunch.

As can be noticed from this example, the rules of the game were not changed. The mine could have denied the community the help they requested and continued working as if nothing had happened (hoping for no negative externalities). But, as mentioned before, the community and corporate strategies and goals are mutually beneficial, which is why the mine agreed to help, providing the community the extra financial help they needed and requested.

The Central American countries require significant public investment to meet the SDGs, probably beyond their current capacities, to be able to improve the quality of life of their people, and the metal mining companies have demonstrated their willingness and commitment in taking their share of the contribution to economic growth and the overall well-being of their neighbouring communities and hosting countries.

In conclusion, Central America has the presence of the metal mining companies in different and remote parts of the countries. The metal mining companies have the willingness and the resources to help and support the low-income countries to improve people's living conditions, and the Central American countries have the natural resources needed to attract more foreign investors. Governments, communities and mining stakeholders should generate the necessary synergies to connect the interests of both parties, and in this way narrow or eliminate the gap that exists between the low-income countries and their meeting the SDGs.

## REFERENCES

Cambridge Dictionary. (n.d.). Retrieved from Diccionario https://dictionary.cambridge.org/es/diccionario/ingles/latrine?q=latrines.

Guatemala, G. d. (n.d.). Reglamento de la Ley Minera. Retrieved from Decreto 48-97: http://transparenciaold.minfin.gob.gt/transparencia/BibliotecaDigital/Documentos%20%20Area%20Económico/rminaria.pdf.

Guatemala, I. N. (n.d.). INE. Retrieved from Estadisticas por tema: https://www.ine.gob.gt/ine/portal-estadistico-1-0/.

Ianrna-URL. (2012). Perfil Ambiental de Guatemala 2010–2012; Vulnerabilidad local y creciente construccion de riesgo.

Minerals, A. (n.d.). Corporate responsability. Retrieved from http://auraminerals.com/community/our-commitment/.

Nations (n.d.). UNWATER. Retrieved from UNWATER: https://www.unwater.org/water-facts/water-sanitation-and-hygiene/.

Newmont. (n.d.). Newmont's strategy. Retrieved from https://www.newmont.com/about/strategy/.

Newmont-Goldcorp. (n.d.). Mina Marlin. Retrieved from Infraestructura: https://newmontgoldcorp-marlin.com/valor-sostenible/infraestructura/.

Onil Banerjee, J. M. (2019). Evaluating synergies and trade-offs in achieving the SDGs of zero hunger and clean water and sanitation: An application of the IEEM platform to guatemala. *Ecological Economics*, April.

Segeplan. (2011, mayo). Segeplan. Retrieved from http://www.segeplan.gob.gt/downloads/clearinghouse/politicas_publicas/Recursos%20Naturales/Pol%C3%ADtica%20Nacional%20del%20Agua%20de%20Guatemala.pdf.

Silver, P. (2018). Sustainability report 2018. Retrieved from https://www.panamericansilver.com/assets/Sustainability-reports/c176f42888/2018-Sustainability-Report.pdf.

Silver, P. A. (n.d.). United Nations SDGs. Retrieved from https://panamericansilver.com/sustainabilityreport2018/our-goals-and-performance/united-nations-sdgs/.

The Free. Encyclopedia. (n.d.). Retrieved from Wikipedia: https://en.wikipedia.org/wiki/Central_America#Geography.

Unicef, W. (n.d.). JMP. Retrieved from Joint Monitor Program: https://washdata.org/data/household#!/.

Unidas, N. (2017, May 17). Sesion especial sobre los medios de implementacion de la Agenda 2030. Retrieved from https://www.youtube.com/watch?v=7yaAB70zNuw.

United Nations (n.d.). Sustainable development goals. Retrieved from Clean Water and Sanitation: https://www.un.org/sustainabledevelopment/water-and-sanitation/.

United Nations (n.d.). Sustainable development goals. Retrieved from United Nations Development Programme: https://www.undp.org/content/undp/en/home/sustainable-development-goals.html.

Valencia, U. P. (n.d.). Edx. Retrieved from ODS en la Agenda 2030 de las Naciones Unidas. Retos de los Objetivos de Desarrollo Sostenible: https://www.edx.org/es/course/ods-en-la-agenda-2030-de-las-naciones-unidas-retos*.

# GOAL 7 Ensure Access to Affordable, Reliable, Sustainable and Modern Energy for All

*Cristian Parra, Joseph Kirschke and Saleem H. Ali*

## CONTENTS



## INTRODUCTION

Effectively connecting SDG 7 to the extractive industries will require us to move beyond the classic affordability and dependable supply approach to energy security (Pascual & Elkind, 2009). The development imperative is "energy poverty" – particularly providing reliable electricity to much of the developing world which is still highly deficient in access (Aklin et al., 2018). Relieving the burden of energy poverty will require wide-ranging energy infrastructure which in many cases is dependent on metals for both generation and delivery. Furthermore, the extraction process itself for such metals for such infrastructure and for broader usage can also be made more energy efficient.

This chapter presents both facets of this linkage between SDG 7 and the extractive industries. We also provide a deep-dive comparative of Europe and Asia-Pacific in the context of Agenda 2030, the Sustainable Development Goals and specifically SDG 7. The analysis is supported by energy data (Enerdata, 2018) to understand the global and regional tendencies; regional indicators proposed by the Sustainable Development Goals (specifically SDG 7) (UN, 2018) and the ambitious determination of the Paris Agreement (UNFCCC, 2015) to combat climate change. The quantity and quality of demanded changes, as exemplified by Asia and Europe respectively, have important implications for how the extractive industries need to engage with both the developed and developing worlds in their SDG targets.

First, let us consider the energy usage by the extractive industries globally, which is estimated to be around 11% of total energy consumption (Guilbaud, 2016). While this estimate is in constant flux due to the ebbs and flows of mining projects as well as efficiency improvements, there is little doubt that extraction is highly energy intensive for a variety of reasons. Initial energy usage in building infrastructure for projects, especially in remote areas, coupled with the use of transport and hauling equipment in remote parts of the world, has a major energy footprint. However, the most significant operational energy footprint of mining is smelting operations. Smelters are often situated in regions with cheap and reliable electricity supply and present an important opportunity for development in remote areas, such as the Edea aluminium smelter in Cameroon.

It is in this regard that mining also has the potential to provide vital SDG 7 target attainment. As mines electrify worldwide, they can develop infrastructure corridors and generation capacity to serve some of the most energy-poor parts of the world. Figure 7.1 shows the new electricity capacity that is planned up to 2040. In a survey of the mining industry conducted by Ernst and Young (EY) in March 2019, there was a clear linkage made by the respondents between using the electrification of mines to improve development outcomes and also obtaining greater "social license to operate" (Ernst and Young - EY, 2019).

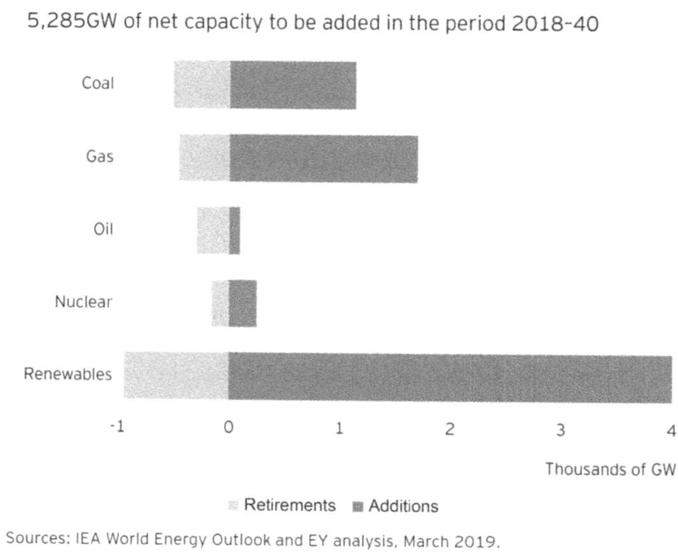


**FIGURE 7.1** Energy consumption by region (Mtoe).

## COMPARATIVE ANALYSIS OF TWO REGIONS AND POTENTIAL FOR EXTRACTIVE INDUSTRIES INTERFACE

The world total energy consumption grew from 8,761 Mtoe in 1990 to 14,126 Mtoe in 2017. Around 70% of this change was driven by the energy consumption of the Asia-Pacific region, which increased its world participation from 25% (1990) to 41% (2017). This energy consumption dynamic of the Asia-Pacific region contrasts with the energy consumption of Europe which has remained at similar levels during the same period. Hence, the consumption ratio between Asia-Pacific and Europe rose from 1.2 (1990) to 3.2 (2017) (consumption of Asia-Pacific for each Europe's consumption unit).

What this comparison misses is that Europe is aiming to move ambitiously towards cleaner delivery of energy and even though total consumption may not grow, the quality of energy delivery in terms of its pollution reduction will likely improve. This is an important area for the mining industry to contribute towards in terms of ensuring the metals to build this new infrastructure of cleaner energy delivery (Table 7.1).

## ACCESS TO AFFORDABLE AND CLEAN ENERGY ASIA-PACIFIC: SDG 7

The access of the population to energy by affordable, reliable and modern services is considered a fundamental sustainable development goal. During the last two decades, while developed countries, more industrialized economies and urban areas of developing economies have expanded the necessary infrastructure for supplying energy services, significant gaps still remain in rural regions and less developed economics, and in terms of energy efficiency and clean supply.

Hence, official data of access to electricity of recent years for China, Malaysia, South Korea and Malaysia shows how these economies have reach 100% of access in

**TABLE 7.1**
**Energy Consumption by Region (Mtoe)**

| | 1990 | 2000 | 2010 | 2017 | Change (2010–2017) | Change (2010–2017) % |
|---|---|---|---|---|---|---|
| Asia-Pacific | 2,213 | 3,014 | 4,971 | 5,909 | **2,895** | **70** |
| Europe | 1,786 | 1,854 | 1,931 | 1,857 | **3** | **0** |
| Other | 4,762 | 5,148 | 5,964 | 6,360 | **1,212** | **29** |
| **World** | **8,761** | **10,016** | **12,866** | **14,126** | **4,110** | **100** |
| Asia-Pacific (%) | 25% | | | 41% | | |
| Europe (%) | 20% | | | 13% | | |
| Asia-Pacific/Europe (Consumption Ratio) | 1.2 | | | 3.2 | | |

*Note:* Prepared by the authors based on ENERDATA statistics.

rural and urban areas (similar to Australia and New Zealand), evidencing impressive progress and how effective their public policies have been.

The challenges linked to universal access in Asia-Pacific remain in rural areas (Table 7.2).

## SDG 7 CHALLENGES IN THE CONTEXT OF THE ENERGY AND RESOURCE SECTOR

According to the 2017 report of The World Energy Council, Europe shows strong performance on energy from the perspective of sustainability and equity (access and affordability); however, there are important long-term challenges to maintain this level of efficiency and achieve a better level of energy security. In order to face the challenges, the report highlights the necessity of more market efficiency and more integration of the European market, national regulations and coordination in subregional policy-making (Table 7.3).

Asia-Pacific, despite its economic, demographic, cultural and geographic diversity, faces common challenges linked to growing energy demands and how to continue exploring different options to reduce greenhouse emissions.

## PROGRESS WORLDWIDE IN CLEAN ENERGY DELIVERY IN THE MINING INDUSTRY

Since 2011, the industry first began economizing with solar and wind. This was just before the crash of a 13-year emerging-market commodities boom which sent people, equipment and infrastructure deep into some of the most far-flung corners of the world scouring for metals and minerals. By 2022, Navigant, a consultancy, forecasts the use of clean energy by miners will grow to 8 percent – or $3.9 billion – from 1.8 percent today.

This month, IAMGOLD, a mid-sized Canadian firm, made news by signing a $20 million, 15-year power purchase agreement for a solar installation at its 15-megawatt power (MWp) off-grid Essakane mine in Burkina Faso. Elsewhere in Africa, German-based Redavia Solar, the largest ground-mount solar company in Tanzania, announced its solar photovoltaic (PV) expansion at Shanta Mining Company Ltd.'s New Luika gold asset. The 609 kilowatt (Kw) installation will yield 950 Kw hours (Kwh) annually; on completion, the entire project will generate 1,040,000 Kwh each year.

Thanks to strong domestic and international support, India is witnessing a clean-energy renaissance for communities, businesses and industries – and mining is no exception. Key participants include coal firms like state-run Coal India Ltd., which plans to install 600 Mw of solar nationwide, along with Neyveli Lignite Corp., which tendered for 260 Mw of grid-connected solar PV in two states. On December 1, Hindustan Zinc joined the group by announcing the solar deployment of 115 Mw; the company already generates 474 Mw of thermal power and 274 Mw of wind.

More than 16,000 miles away, Chile, one of the world's greenest economies and its largest copper producer, is getting a huge clean-energy boost from its red metal mines. As elsewhere, plummeting solar prices combined with technological

**TABLE 7.2**

**Access to Electricity Asia-Pacific (1)**

| | % of Total Rural Population with Access to Electricity | | | | Rural Population | | |
|---|---|---|---|---|---|---|---|
| | 1990 | 2000 | 2010 | 2016 | Rural Population as % of Total Population | Total Rural Population (Thousand) | Total Rural Population without Access to Electricity (Thousand) |
| Mongolia | 25.8 | 25.1 | 16.8 | 44.2 | 31 | 953 | 532 |
| Indonesia | 48.4 | 79.7 | 89.4 | 94.8 | 45 | 118,796 | 6,209 |
| Lao PDR | 0.7 | 28.3 | 57.2 | 80.3 | 65 | 4,458 | 877 |
| Cambodia | | 9.0 | 18.8 | 36.5 | 77 | 12,320 | 7,828 |
| Myanmar | 20.3 | 28.9 | 34.3 | 39.8 | 69 | 36,825 | 22,172 |
| Philippines | 40.1 | 58.5 | 75.8 | 86.3 | 53 | 55,607 | 7,641 |
| Timor-Leste | | 2.9 | 24.0 | 49.2 | 70.0 | 907 | 461 |
| Papua New Guinea | | 4.4 | 9.0 | 15.5 | 86.0 | 7,096 | 5,998 |
| Pacific Islands | | | | 71.0 | 61 | 1,476 | 428 |
| India | 29.7 | 48.2 | 65.7 | 77.6 | 66 | 884,400 | 197,852 |
| Pakistan | | | | 98.8 | 63 | 124,110 | 1,480 |
| Bangladesh | | | | 68.9 | 64 | 104,960 | 32,695 |
| Total Rural Population Asia-Pacific without Access to Electricity | | | | | | | **284,173** |

*Source:* The World Bank – Development Indicators.

**TABLE 7.3**
**Challenges and Opportunities in the Context of SDG 7 and Agenda 2030 (Prepared by the Authors Using Trilemma Framework of WEC)**

| | Challenges | | Opportunities for Cooperation with Extractive Industries (Target 7.a and 7.b) |
|---|---|---|---|
| | Europe | Asia-Pacific | |
| Security (Target 7.1) (Target 7.2) (Target 7.3) | • Production efficiency<br>• Innovation<br>• Grid efficiency<br>• Regulation and market integration<br>• Cooperation at subnational level for policy-making | • Growing demand<br>• Policy and geopolitical approach transition from "security of source" to long-term collaboration approach.<br>• Investment challenges<br>• Legal and political barrier for energy investment | • Policy design cooperation<br>• R&D<br>• Collaboration for clean technology development: solar panels, batteries |
| Equity (Target 7.2) | • Pricing for specific sectors | • Access to electricity in rural areas<br>• Growing demand in urban areas | • Partnership and multi-stakeholders' cooperation to energy access in rural areas<br>• Public-private partnership |
| Sustainable (Target 7.2) (Target 7.3) | • Finance for renewable energy<br>• New green technologies | • Reduce CO2 Emission<br>• Green energy investment based on partnership for development.<br>• Liabilities of developers | • China: To complete national emission trading scheme by 2020 (trading platform) |

*Notes:*

7.1 Ensure universal access to affordable, reliable and modern energy services

7.2 Increase substantially the share of renewable energy in the global energy mix

7.3 Double the global rate of improvement in energy efficiency

7.a Enhance international cooperation to facilitate access to clean energy research and technology, including renewable energy, energy efficiency and advanced and cleaner fossil-fuel technology, and promote investment in energy infrastructure and clean energy technology

7.b Expand infrastructure and upgrade technology for supplying modern and sustainable energy services for all in developing countries, in particular least developed countries, small island developing States and landlocked developing countries, in accordance with their respective programmes of support.

advances mean that – from the sun-baked Andes Mountains thousands of feet above sea level to blustery southern Patagonia – solar panels and wind turbines are fast replacing the need to truck diesel fuel to these harsh, distant places – a costly and unreliable enterprise.

In December 2019, for example, reports emerged that interest among Chile's copper miners in renewables has ballooned so quickly that traditional energy suppliers must now diversify their portfolios. By July 2020, Chile further planned to install 1,000 Mw of solar – mostly in mining regions – while Chile's Association for

Renewable Energy has projected 100 percent of the national grid could be powered by renewables as of 2050.

Meanwhile, a 1,900-mile transmission line is being constructed to expand the power distribution system by connecting mines to the national grid. This has hemispheric implications: Mining already accounts for 60 percent of Latin America's solar PV market, while the industry is expected to represent nearly two-thirds of energy usage by the region's top 21 solar plants.

In other parts of the world, the subsector has shown myriad benefits, too. These include shareholder demands for climate change transparency and – especially – the bedrock community relations, or "social license to operate," essential to any mining operation. These installations create jobs while unlocking revenue for mining companies to invest in local schools, agriculture and medical clinics along with small- and large-scale infrastructure.

IAMGOLD, the largest private-sector employer in Burkina Faso, for instance, created 120 jobs with the initial installation of a 15-Mw solar plant at its West African mine. In South America, the Toronto-based firm also created more than 100 jobs with the deployment of Suriname's first and largest solar farm with 5 Mw at its Rosebel gold asset. In all, IAMGOLD sought to use 15 percent renewable energy company-wide by 2020.

Many such solar installations will be left behind for local use once the mine is closed. These also open the door for "anchor-tenant solutions," whereby operating mines share electricity with far-off local communities. This is important for places like Africa, where 80 percent of new mines will be clean-energy-dependent by 2026, according to the International Finance Corp., and where 600 million people live without electricity.

Global institutions have taken note. The World Bank, in its 2015 report "Power of the Mine: A Transformative Opportunity for Sub-Saharan Africa," for instance, notes mining's appetite for power in sub-Saharan Africa will triple between 2000 and 2020 to reach more than 23,000 Mw.

One of the most significant examples lies in Queensland, Australia, where Anglo-Australian multinational Rio Tinto operates a bauxite mine with an 18,000-panel PV farm installed by Arizona-based First Solar, which shares electricity with a nearby township. Rio, the world's second-largest mining company, has recently announced plans to expand the project.

Also of major importance is the prospect of further greening some of the world's largest industrial-scale infrastructure projects associated with mining. These include "pit to port" capacities ranging from smelters, water and other processing systems to conveyor belts, roads and railways along with port and shipping facilities.

Given that 90 percent of energy demand will come from emerging markets by 2035, requiring $1 trillion in investments, according to the International Energy Agency (IEA), the potential synergies are endless. The UN's "Sustainable Energy for All" (See4All) initiative, for example, seeks to create universal energy access by 2030, while Power Africa, a U.S.-led public-private partnership, has sought to double energy access to 60 million, partly through renewables, across the continent.

Thus the mining industry, while reforming its own energy usage through cleaner production also has the potential to deliver electricity to underserviced parts of the world, thereby reducing energy poverty and meeting many of the salient targets of SDG 7.

## REFERENCES

Aklin, M., Harish, S.P., and Urpelainen, J. (2018). A Global Analysis of Progress in Household Electrification. Retrieved from https://www.sciencedirect.com/science/article/abs/pii/S030142151830466X.

Department of Economic and Social Affairs. Retrieved from https://sustainabledevelopment.un.org/sdg7.

Enerdata. (2018). Global Energy Statistical Yearbook 2018. Retrieved 2019, from Enerdata https://www.enerdata.net/publications/world-energy-statistics-supply-and-demand.html.

Ernst and Young (EY). (2019). Will Electrification Spark the Next Wave of Mining Evolution. Retrieved from https://www.ey.com/en_us/mining-metals/will-electrification-spark-the-next-wave-of-mining-innovation.

Guilbaud, J.J.S. (2016). Hybrid Renewable Power Systems for the Mining Industry: System Costs, Reliability Costs, and Portfolio Cost Risks. Doctoral thesis. UCL (University College London).

Pascual, C., and Elkind, J. (2009). *Energy Security: Economics, Politics, Strategies, and Implications*. Washington, DC: Brookings Institution Press.

United Nations. (2015). Transforming Our World: The 2030 Agenda for Sustainable Development. In: *Resolution Adopted by the General Assembly on 25 September 2015* (1st ed., pp. 35). New York: United Nations.

United Nations. (2018). Sustainable Development Goals. Retrieved United Nations 2019, from Division for Sustainable Development Goals.

UNFCCC. (2015). Re: Paris Agreement. Message Posted to https://unfccc.int/resource/bigpicture/#content-the-paris-agreemen.

WEC. (2018). World Energy Trilemma Index 2018. Retrieved from London https://www.worldenergy.org/.

# GOAL 8 Promote Sustained, Inclusive and Sustainable Economic Growth, Full and Productive Employment and Decent Work for All

*Gavin Hilson*

## CONTENTS



## INTRODUCTION

Media coverage which demonizes artisanal and small-scale mining (ASM) – low-tech, labour-intensive mineral extraction and processing – is growing. Negative opinions of the sector's activities and operators are deeply ingrained and have influenced public opinion worldwide. They have, more importantly, made it challenging to convince experts of ASM's importance and the need for it to feature more prominently in economic and development strategies. This is despite a growing body of literature which highlights in considerable depth ASM's economic impact worldwide, beginning with its ability to create jobs in even the most impoverished of settings. Worldwide, there are at least 20 to 30 million people employed directly in the sector (Buxton, 2013), in sharp contrast to large-scale mining which, because of its capital-intensive and enclave nature, creates comparatively far fewer jobs. Where data are available, the discrepancy between the two in the area of employment is sizable: in Tanzania, at least 650,000 engaged in ASM versus 8800 in large-scale mining; in Mongolia, 100,000 versus 23,000; in Ghana, 1.5 million versus 11,628 (according to the latest figures supplied by the country's 12 producing mines); and in the Philippines, over 300,000 versus the combined 7300 staff of the country's major mine operators, including SR Metals and Carmen Copper Corporation (Buxton, 2013; Cielo, 2015; O'Faircheallaigh and Corbett, 2016; Muff et al., 2016; Corbett et al., 2017; Moore, 2018; Schoneveld et al., 2018). This body of literature also highlights how in many sections of Latin America, sub-Saharan Africa and Asia, ASM dovetails and injects revenue into agriculture, accounts for a sizable share of nonfuel mineral output, and stimulates the growth of numerous downstream and upstream industries.

These *positives* are routinely dismissed and lost in discussions which point to how – in most cases, exaggeratedly – the sector is environmentally destructive, its labour force faces health and safety concerns and its workers toil in precarious conditions. Public perception of ASM must, therefore, radically change if the sector's benefits are to be fully realized. It begins with views on the sector's labour practices: people are so quick to portray the work conditions and operations of miners negatively without giving much consideration to the context which gives rise to activities in the first place. But how can this realistically be achieved when donors, host governments and even certain NGOs appear to have very little interest in changing this perception, and continue to downplay the dynamic role ASM plays across the developing world?

The international development policy machinery, consequently, continues to evolve with minimal appreciation of ASM's economic importance in impoverished sections of Asia, Latin America and sub-Saharan Africa. The origins of this crucial oversight can be traced back to the late 1980s and early 1990s, a time when most mineral-rich developing countries were in the midst of resuscitating underperforming mineral exploration facilities and reforming their large-scale mining economies, many under the tutelage of the World Bank and the European Commission. Anticipating that the sweeping changes being made to mineral policies at this time would catalyze an influx of foreign investment, Davidson (1993) called for a ring-fencing of territory and commitment of resources needed to properly formalize and support ASM:

> Governments must be prepared to move beyond the establishment of legal frameworks, to identify deposits and areas amenable to small-scale development, including the preliminary evaluation of their technical and economic viability at different levels of operation. Security of tenure should be respected in such areas. [p. 317]

The author's choice of the word 'must' is worth highlighting here because most governments did precisely the opposite: they ignored ASM. Whilst there were provisions to formalize and support ASM built into the financial packages awarded by the World Bank and EU, it was clear, at an early stage, that this was not going to be a priority undertaking. It was rather a distant second goal to the principal objective of attracting investment to develop further capital-intensive large-scale mineral exploration and extraction, which has required committing vast tracts of land to support activities.

Perhaps the haste with which reform of large-scale mining and mineral exploration economies was generally carried out was due to the donor community failing to recognize fully the economic importance of ASM. This would come as no surprise because aside from analysis presented by Davidson (1993) and a small group of other scholars and experts (e.g. Barry, 1996; Labonne, 1996), the subject (i.e. ASM's economic impact) received very little coverage in the literature at the time. It was not until the turn of the century that the livelihoods dimension of ASM began to capture the attention of scholars, a time when a range of dynamic sector-specific interventions themed around subjects such as gender, child labour and supply chains were implemented by donors. But with mining and development strategies and donor landscapes having changed so much by this time, shaped heavily by entrenched policy with a distinctive large-scale 'bias', any project implemented to support ASM, when 'left to stand on its own, … often gently wound down due to a lack of continued government support or supervision' (ILO, 1999, np).

The failure to properly integrate ASM into broader rural development and poverty alleviation strategies in sub-Saharan Africa, Asia and Latin America has proved to be an even bigger oversight. It began with the Millennium Development Goals (MDGs), the architects of which had a very specific agenda and vision for facilitating economic growth in Asia, sub-Saharan Africa and Latin America. On the subject of rural poverty alleviation, the MDGs were extremely formulaic, stressing heavily the need to support subsistence agriculture – despite its obvious limitations as development strategy – which spawned very specific interventions, such as the Millennium Villages in sub-Saharan Africa. With ASM having gone virtually overlooked during the numerous dialogues which took place prior to the launch of the Millennium Project, experts were forced to weigh in, retrospectively and rather aimlessly, on how the sector, if properly supported, could help to address targets linked to the Goals (see e.g. Hobbs, 2005; World Bank, 2005).

There was an opportunity to shift gears with the Sustainable Development Goals (SDGs). At the time of their launch, significant research had been conducted across the developing world which captured in detail ASM's economic impact (Dreschler, 2001; Hentschel et al., 2002; Hilson and Garforth, 2012). Relevant findings had been disseminated at successive UN Commission on Sustainable Development sessions, which spawned vibrant discussion about the sector's importance in countless rural and peri-urban communities across the developing world. But much like the MDGs before them, the SDGs would also end up being designed without ASM in mind. An opportunity arose post-launch to partly address this oversight by showcasing ASM and highlighting the importance of making it a centrepiece of national extractive industries-led development policies in the *Mapping Mining to the Sustainable Development Goals: An Atlas* (United Nations, 2016). Its authors, however, elected not to do so. They rather decided to focus solely on large-scale mining in the *Atlas*, on the grounds that, in their view, 'The scale of ASM warrants a separate guide and review to map the opportunities on how ASM can contribute to the SDGs' (p. 15). The decision reaffirmed what is already known: how, despite growing recognition of ASM's livelihoods dimension, policymakers and to some extent, donors, remain content with it continuing to be regulated, managed and supported in isolation – on the fringes of the international and regional development agenda.

The challenge now, in light of these missed opportunities, is finding creative ways to showcase the sector's economic importance but which closely align with the broader international development policy machinery. This is precisely what this chapter attempts to achieve by devising a blueprint to facilitate 'decent work and economic growth' for ASM which aligns with SDG 8 (See Table 8.1 for a list of targets linked to SDG 8). In doing so, the chapter reflects on ways in which such a blueprint can be grafted on to and, very importantly, have some traction within, the

**TABLE 8.1**
**SDG 8 and Accompanying Targets**

| Target | Details |
|---|---|
| Target 8.1 | Sustain per capita economic growth in accordance with national circumstances and, in particular, at least 7 per cent gross domestic product growth per annum in the least developed countries. |
| Target 8.2 | Achieve higher levels of economic productivity through diversification, technological upgrading and innovation, including through a focus on high value-added and labour-intensive sectors. |
| Target 8.3 | Promote development-oriented policies that support productive activities, decent job creation, entrepreneurship, creativity and innovation, and encourage the formalization and growth of micro-, small- and medium-sized enterprises, including through access to financial services. |
| Target 8.4 | Improve progressively, through 2030, global resource efficiency in consumption and production and endeavour to decouple economic growth from environmental degradation, in accordance with the ten-year framework of programmes on sustainable consumption and production, with developed countries taking the lead. |
| Target 8.5 | By 2030, achieve full and productive employment and decent work for all women and men, including for young people and persons with disabilities, and equal pay for work of equal value. |
| Target 8.6 | By 2020, substantially reduce the proportion of youth not in employment, education or training. |
| Target 8.7 | Take immediate and effective measures to eradicate forced labour, end modern slavery and human trafficking and secure the prohibition and elimination of the worst forms of child labour, including recruitment and use of child soldiers, and by 2025 end child labour in all its forms. |
| Target 8.8 | Protect labour rights and promote safe and secure working environments for all workers, including migrant workers, in particular women migrants, and those in precarious employment. |
| Target 8.9 | By 2030, devise and implement policies to promote sustainable tourism that creates jobs and promotes local culture and products. |
| Target 8.10 | Strengthen the capacity of domestic financial institutions to encourage and expand access to banking, insurance and financial services for all. |
| Target 8.a | Increase Aid for Trade support for developing countries, in particular least developed countries, including through the Enhanced Integrated Framework for Trade-Related Technical Assistance to Least Developed Countries. |
| Target 8.b | By 2020, develop and operationalize a global strategy for youth employment and implement the Global Jobs Pact of the International Labour Organization. |

SDGs framework and broader development policy machinery. Ironically, despite this oversight, the formula for facilitating 'decent work and economic growth' in ASM is captured within SDG Targets 8.2 and 8.3. Changes made to ASM policy frameworks which speak to these targets would enable governments to address Target 8.7, and ultimately lay the groundwork for a blueprint for 'decent work and economic growth'.

## CONFUSION OVER CERTIFICATION AND STANDARDS

The lack of policy guidance and support for ASM, and the absence of a referenceable document that outlines specifically how supporting the sector can contribute to the SDGs, has led many organizations, scholars and experts to present their own views on what constitutes the 'ideal' (ASM) operation. Few engage with the SDGs explicitly but do take the liberty of presenting what they believe to be a range of standards, frameworks and/or guidelines capable of facilitating marked improvements in the sector. This is perhaps best epitomized by the flurry of recent efforts to devise standards for ethical minerals. Fairmined Gold, for example, is an intervention which its architects claim was conceived specifically for 'creating opportunities for artisanal and small-scale miners and their communities, not medium and large industrial mining'.* The designers of the Fairtrade Gold Standard appear determined to achieve the same, claiming that 'Fairtrade's vision for artisanal and small-scale mining is aligned to the core Fairtrade values of empowering producers and local communities through trade and delivering economic, social and environmental transformation and restoration'.†

The problem, however, is that the trajectories and accreditation criteria linked to both interventions and most other mineral certification schemes are untenable for over 90 per cent of ASM operators worldwide. As has been widely discussed in the literature (ILO, 1999; Hentschel et al., 2002; Hilson et al., 2018), a combination of bureaucratic regulations, costly registration processes and a lack of available land on which to place ASM licensees has pushed operators into the informal economy, in most cases, confining them to this existence. Referring back to Target 8.3, therefore, a necessary first move is the establishment of a bedrock capable of facilitating innovation and growth in the ASM sector – which formalization provides. With ASM providing tens of millions of people with direct employment and creating hundreds of millions of additional income-earning opportunities in downstream and upstream activities in 80 countries worldwide,‡ the case for formalizing the sector on economic grounds is clear. Formalization in this context, however, is more than a streamlined

* 'FAIRMINED STANDARD FOR GOLD FROM ARTISANAL AND SMALL-SCALE MINING, INCLUDING ASSOCIATED PRECIOUS METALS', Alliance for Responsible Mining, Bogota www.fairmined.org/the-fairmined-standard/ (Accessed 4 January 2020).

† 'Fairtrade Standard for Gold and Associated Precious Metals for Artisanal and Small-Scale Mining', Fairtrade International, https://files.fairtrade.net/standards/2015-04-15_EN_Gold-and-Precious_Metals.pdf (Accessed 4 January 2020).

‡ Estimate based on several donors' projections, including those of the World Bank and the International Labour Organization. This is the author's projection in the absence of any tangible data.

and simplified licensing system. It is rather a series of processes and establishment of a policy framework within which a foundation is provided for registered operators to 'grow' and position themselves to develop their activities in ways they have been unable to do previously:

> Formalization is a process that seeks to integrate ASGM [artisanal and small-scale gold mining] into the formal economy. The process of formalization includes the development or adaptation of mining (and other) laws or policies to address the challenges of ASGM.
>
> **[UNEP, 2012, p. 1]**

With these, and similar, definitions, the implication is that formalization yields more efficient, regulatable practices. But the concept of formalization is routinely used cavalierly, in many cases being equated with *mechanization*. The root of this problem can be traced back to comments voiced in the landmark report, *Social and Labour Issues in Small-Scale Mining* (ILO, 1999), published by the International Labour Organization (ILO) over two decades ago. The document's authors pointed out how 'Many projects to assist small-scale mining have failed or have not led to lasting improvements because they have treated small-scale mining as a subset of large, formal mining' (np). Perhaps at the time, it did not seem to be a serious enough issue to address but, in hindsight, the way in which operational standards, frameworks and guidelines have been developed for ASM, it is clear that a revisiting of what are often branded as archaic debates around definitions of the sector was a necessary intervention. The question of 'What is ASM?' dominated discussions at inaugural international meetings about the sector, organized in the 1980s and early 1990s, which at the time seemed to agitate some experts (e.g. Hollaway, 1991; Jennings, 2003), who would subsequently comment on how these events were unhelpful and a waste of precious time.

But with *formalization* in this context now being routinely used synonymously with *mechanization* and *upscaling*, a revisiting of debates on how to define ASM seems essential. What is most surprising is how detached the architects of the suite of standards, frameworks and guidelines around labour practices, health and safety concerns and the environment in ASM have been from the burgeoning body of analysis on the sector's organizational structures and the types of people who pursue work and remain involved for lengthy periods at sites. Even a superficial level of engagement with this literature would reveal quite quickly how eclectic ASM populations are, and the magnitude of difference between the livelihood trajectories of the many groups working in the sector. At one extreme of the ASM labour spectrum, there are those who aspire to mechanize. These individuals are typically armed with expertise and ample working capital. Those who fall into this category of miner comprise but a minute fraction of the global ASM workforce. They are primed to mechanize and 'grow' their businesses.

At the other extreme are many more individuals whose association with ASM is often temporary: the students who are mining solely to pay their tuition, the single mother who needs to mine to earn enough money to buy fertilizers for the family farm, and the struggling rural family which needs an injection of finance to support their struggling taxi or chemical dispensary business. The example of Papua New Guinea, where miners have gone on record claiming that 'The river is our ATM!' (Jenkins, 2016, p. 198), in reference to how they turn to local waterways to gather alluvial gold when in need of money, seems to apply to this group of miners, none of whom really aspire to mechanize and acquire heavy machinery. They rather turn to ASM out of desperation, eking out a living through subsistence activity. They would stand to benefit from formalization as well but very differently: it would provide them with the security of tenure needed to carry out their work.

The same applies to the tens of millions of individuals who are sandwiched between these two extremes. They are ASM's equivalent of the 'missing middle': the neglected small business owner or entrepreneur who does not wish to continue operating informally but who is unlikely to achieve medium-scale status. Each has his/her own reasons to engage in ASM and covets the security of tenure a licence or permit affords because it allows them to mine fairly freely and innovate incrementally. But these miners still operate at a fairly artisanal scale, and could never accumulate the capital needed to acquire the heavy machinery that could replace the manual labour they depend upon.

Significantly, all of these examples fall squarely into the category of ASM: despite the visible differences between the two extremes and the variations in the activities which fall in between, every operation along this spectrum is an example of low-tech, labour-intensive mineral extraction and processing in its own way. But these differences become significant and must be accounted for when work practices, standards and guidelines are devised for the sector. As the next section of the chapter explains, formalization provides a much-needed platform for facilitating improvements in work practices and stimulating economic development in the sector. As noted, formalization in this context must be viewed as a series of *processes* or a *process* and be sufficiently dynamic to speak to the full gamut of livelihood trajectories associated with ASM.

## FORMALIZATION: NOT THE 'SILVER BULLET' BUT CERTAINLY A STEP IN THE RIGHT DIRECTION

As indicated, several guidelines, frameworks and standards have emerged for ASM over the past two decades. Each was designed with the intention of serving as a blueprint for facilitating improved work standards and more efficient production in the sector. But without a comprehensive understanding of the structures of ASM, the origins of the different people found working at sites, the relationships they have forged with one another and other actors, and the drivers behind the sector's growth, the architects of these schemes have ended up putting together designs which cater almost exclusively to the small group of operators who are armed with capital and have the capacity to innovate unimpeded. Referring once again to the ASM spectrum, these are typically the individuals engaged in mechanized activities. To avoid devising yet another set of inapplicable guidelines for ASM and in the process, perpetuating this problem, it is imperative that standards and strategies for the sector are malleable – that is, applicable to a range of contexts.

The obvious starting point with any ASM formalization exercise is ensuring that all groups of individuals – affluent and subsistent – are able to access the licences and permits that provide them with security of tenure. From accounts in the literature, there are very specific ways in which this can be achieved, including blocking out areas for prospective licensees, decentralizing the decision-making processes on applications and minimizing the costs for registration (ILO, 1999; Hentschel et al., 2002; Hilson et al., 2018). There is the added challenge of how to package ASM formalization more holistically and in ways which resonate with donors and governments, given the lack of traction the sector has in the international development agenda. Although again, the SDGs were designed with little regard for ASM's importance, they are broad enough and intersect many of the ideas behind the drive to formalize the sector. There are opportunities, therefore, to package and rebrand ASM formalization in more concrete development lingo which resonates more powerfully with decision-makers, and in the process, give this issue greater visibility and more legitimacy (Hilson and Maconachie, 2020). This, it is believed, is the first building block towards addressing Target 8.3.

Once this is accomplished, the second, perhaps more critical, step must be undertaken: formulating the 'process' dimension of formalization alluded to earlier, which this section of the chapter will address in the context of SDG 8. If formalization is to be catalytic and transformational, it must speak to, and provide tangible solutions for, the full suite of ASM operators. Specifically, it must yield a dynamic platform capable of empowering, and which provides autonomy to, individuals engaged in ASM who are on very different work and development trajectories, and provide a much-needed foundation for supporting an array of assistance. All of these elements comprise a *process* which is capable of facilitating improvements across the sector. In addition to being grounded in the realities of ASM, the priority areas identified in the following subsection are themselves connected to different SDGs. The intention is to devise a topical and engaging blueprint which resonates with donors and is aligned with international development strategy.

## Focus on Mercury

Bringing ASM into the legal domain puts governments in a much better position to regulate and monitor all types of activities. This is significant in the context of mercury, which features in most ASM activities worldwide: it is relied upon to amalgamate gold. As the largest anthropogenic source of mercury pollution, ASM routinely gets criticized for its impact on the environment. It is a major reason why the UN launched the *Minamata Convention on Mercury*, which officially took effect in August 2017. Countries that have ratified the Convention with quantities of ASM activity that are 'more than insignificant', must, in line with Article 7, 'take steps to reduce, and where feasible eliminate, the use of mercury and mercury compounds in, and the emissions and releases to the environment of mercury from, such mining and processing'.

Given its wide availability and efficiency, and the fact that there is a dynamic microeconomy populated by thousands of actors facilitating its distribution, mercury is unlikely to ever be phased out of the sector entirely. This point, however, is beyond the scope of the current chapter. If the goal is to 'phase out' and find 'alternatives', as the Convention indicates, then flexible policies must be in place which allow regulators to engage with miners of all types. In countries such as Colombia, for example, once operators are in the legal domain, government officials must be prepared to implement comprehensive plans which will differ markedly from location to location. In the Department of Antioquia, for example, it probably should engage directly with people working with and at the 323 'entables' or processing centres which handle mercury (García et al., 2015). In the Department of Cauca, where surveys revealed that 87.5 per cent of ASM operators have no mining titles and, in 2011 alone, an estimated 15,806 kg of mercury was used in the sector, the approaches taken would need to cater more to operators engaged in more rudimentary activities (Vélez-Torres et al., 2018). In Suriname, a different approach must be pursued altogether. Here, whilst most ASM is informal due to it being loosely regulated by government, the sector is populated by both local Maroons and migrants (typically, Brazilians). The sector also has a sizable contingent of mechanized activities and on the whole, the workforce is fused, with Maroons and migrants working together as teams (Seccatore and de Theije, 2017). It will likely be more challenging to engage miners in Suriname, where most are widely dispersed in the country's interior, than say neighbouring Guyana, a country with ASM activities populated by elites who control the industry's national association and have significant political influence (Pasha et al., 2017).

Regardless of any programme or plan to tackle mercury, formalization makes this possible. It provides the platform to deviate from the *status quo* and to implement a comprehensive strategy capable of reaching the full spectrum of ASM operators.

## Focus on Work Practices

For decades, there has been a fixation on 'correcting' ASM work practices. The ILO has been instrumental in raising awareness of health and safety conditions at sites through a series of reports published in the 1990s and early 2000s (e.g. ILO, 1999, 2001), which organizations such as Fairtrade, the Alliance for Responsible Mining and OECD* have since drawn even more attention to and developed near-impossible standards for the sector's operators to adhere to if they wish to have their activities certified.

This, however, begs the question: what are the optimal work practices in ASM? By definition, ASM is labour-intensive, which comes with its own challenges. The problem with most of the guidelines around work practices linked to ethical mineral schemes and interventions with roots in the *OECD Due Diligence Guidance* is that they set standards which, for most operators, are unattainable. These moves further

* Specifically, the *OECD Due Diligence Guidance for Responsible Supply Chains of Minerals from Conflict-Affected and High-Risk Areas* (OECD, 2013).

underscore how, without adequate government oversight and data, inappropriate policy interventions continue to be designed for ASM. The *OECD Supplement on Gold*, for example, states, on the one hand, that

> Artisanal and small-scale gold producers such as individuals, informal working groups or communities are not expected to carry out due diligence as recommended in this Guidance, but they are encouraged to remain involved in due diligence efforts of their customers and formalise so they can carry out due diligence in the future.
>
> **(p. 64)**

Yet, on the other hand, it makes this demand, despite acknowledging that 'ASM is normally low capital intensive and uses high labour intensive technologies' (p. 65), a level from which 'to formalise so they can carry out due diligence in the future' is almost impossible without support – assistance that the OECD will not provide on the grounds that the operators in question are informal.

But it is the failure to provide any context on key subjects which is perhaps the biggest shortcoming of frameworks such as the *OECD Due Diligence Guidance* and *Responsible Jewellery Council Standards Guidance* (RJC, 2013). 'Child labour', for instance, is a phrase mentioned 97 times in the latter but without explanation or any appreciation of what leads boys and girls to mine. This oversight is significant in highly impoverished and subsistence locations such as sub-Saharan Africa, where the question of 'what is child labour?' has long been contested because the settings in which young boys and girls are seen working often make the activities they are undertaking appear much worse than they actually are: walking to fetch firewood for the household, collecting water from the river and carrying out various tasks on the family farm (Hilson, 2012). The important takeaway here is that debates on child labour in places such as sub-Saharan Africa transcend any work on this topic in ASM. By failing to acknowledge this, the architects of frameworks such as the *OECD Due Diligence Guidance* and the *Responsible Jewellery Council Standards Guidance* have oversimplified what is a very complex issue.

A commitment to formalization positions governments to connect with populations and to spend the time needed to study and better understand each miner's circumstances. Doing so could, for example, reveal how children present onsite are not engaged in dangerous work as initially believed but rather fetching drinking water and preparing food for their parents; that whilst calls for 'women's conditions to be improved' should be taken seriously, their decision to mine is a conscious one because of their ability to earn ten times what they would make elsewhere; and how the deposit being worked is alluvial, thus rendering all safety criteria devised for underground activity inapplicable. Each situation necessitates the application of very different solutions, and when viewed in the context of SDG 8, it becomes clear how a move made by the concession holder of a semi-mechanized ASM operation to provide safety training to employees and supply them with hardhats and a community-run operation, and how the management of a small alluvial site which puts a stop to women engaging in heavy lifting, both speak equally to Target 8.8.

## Focus on Support

Even though recent funding has been provided for ASM in the form of the UNDP's Development Minerals Program and the World Bank's support for the DELVE Database, there is still a lack of reliable long-term technical and operational investment in the mines themselves. The funding gap is sizable even for licensed miners, very few of whom are able to secure bank loans unless they use assets such as houses, cars and equipment as a down-payment. Much of this has to do with perception, specifically, lenders' lack of trust in the sector. It also has to do with the architects of standards and schemes believing that there is an appetite for alternative livelihoods when, in fact, ASM is, in most cases *the* alternative.

With formalization, however, the gap between operators and government is bridged, positioning the latter to diagnose the financial and technical needs of the former. The needs of the miners entering the formal domain are as diverse as the group itself. There is room, therefore, for the full suite of schemes: from the high-risk lending found in Guyana (Hilson and Laing, 2017), through interventions which mirror the Village Savings and Loans Associations implemented by the NGO IMPACT in DR Congo (IMPACT, 2019), to the equipment-based lending made available for semi-mechanized operators tested in parts of Ghana (Hilson and Ackah-Baidoo, 2011). Formalization in this context should catalyze a cycle, bringing together miners under one umbrella. Doing so would yield improved regulation and monitoring of activities, in turn facilitating training and education for a range of operators in a number of areas, and the collection of data on individuals and their production. A better regulated and taxed ASM sector would provide a much-needed boost of confidence to the bank managers, microfinanciers and equipment providers in the *formal* economy who have the resources to further support the sector.

For miners, these changes would very importantly increase access to the loans, grants and technologies they desperately covet. They would lead to innovation of all types, from the mechanization and transition from mercury to cyanide which the architects of the Africa Mining Vision and the GEF Gold Program* are working to encourage, through to individual panners looking to get their hands on a water pump. This cycle would furthermore facilitate the traceability of flows of capital and minerals in ASM supply chains, which interventions such as the Extractive Industries Transparency Initiative and the Kimberley Process have struggled to achieve. Overall, the cycle of formalization described here and depicted graphically in Figure 8.1 would contribute directly to Target 8.2, Target 8.3 and Target 8.4.

* The Africa Mining Vision is 'a pathway, formulated by African nations themselves, that puts the continent's long term and broad development objectives at the heart of all policy making concerned with mineral extraction' (see 'Africa Mining Vision' www.africaminingvision.org/, accessed 4 January 2020). The GEF Gold Program, which is funded by the Global Environment Facility (GEF) and led by United Nations Environment Programme (UN Environment), aims to help small-scale gold miners in eight countries (Burkina Faso, Colombia, Guyana, Indonesia, Kenya, Mongolia, Peru and the Philippines) replace toxic mercury with cleaner techniques, as well as improve access to finance and facilitate formalization of the sector (see 'Implementation Launch of the GEF GOLD Programme', www.thegef.org/events/implementation-launch-gef-gold-programme, accessed 3 January 2020).

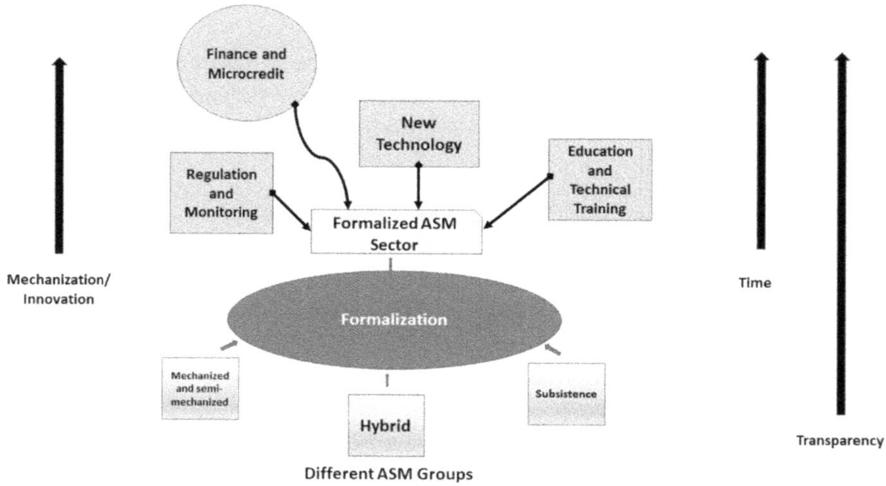


**FIGURE 8.1** Conceptualizing formalization as a catalytic process in the ASM sector.

In summary, by emphasizing actions and improvements in each of the three areas identified, ASM could play an important role in achieving several targets set under SDG 8.

## CONCLUSION

The international development community no longer has the luxury of being able to ignore the ASM sector. But with the SDGs and the MDGs before them having been designed without recognizing ASM's importance, it has proved difficult to build a case for supporting the sector's operators. The former, however, are fairly flexible: if packaged in development terms, strategies which seek to improve the sector and give it the visibility it deserves could gain some traction.

In the case of SDG 8, the starting point is formalization, which, if successful, provides a platform for identifying and addressing more comprehensively the needs of different operators. In recognition of the eclectic collection of activities found in ASM and the need to avoid repeating the mistakes of the past by devising yet another inappropriate framework or set of standards, the chapter highlighted three priority areas for governments: (the management of) mercury, work practices and support. Improvements made in these three areas of the sector would go a long way toward addressing the targets set under SDG 8. The strategies pursued to do so, however, will ultimately vary, depending on the setting.

## REFERENCES

Barry, M. (Ed.) 1996. *Regularizing informal mining. A summary of the proceedings of the international roundtable on artisanal mining.* Organized by the World Bank, Industry and Energy Department Occasional Paper No. 6, Washington, DC.

Buxton, A. 2013. *Responding to the challenge of artisanal and small-scale mining how can knowledge networks help?* International Institute for Environment and Development (IIED), London.

Cielo, M. 2015. *The mining for development framework for the Philippines.* UPSE Discussion Paper No 2015-215. University of the Philippines, Quezon City.

Corbett, T., O'Faircheallaigh, C., Regan, A. 2017. 'Designated areas' and the regulation of artisanal and small-scale mining. *Land Use Policy* 68: 393–401.

Davidson, J. 1993. The transformation and successful development of small-scale mining enterprises in developing countries. *Natural Resources Forum* 17(4): 315–326.

Dreschler, B. 2001. *Small-scale mining and sustainable development within the SADC region.* Mining, Minerals and Sustainable Development (MMSD). International Institute for Environment and Development and World Business Council for Sustainable Development, London.

Hentschel, T., Hruschka, F., Priester, M. 2002. *Global report on artisanal and small-scale mining.* Minerals Mining and Sustainable Development (MMSD) Project. International Institute for Environmental Development, London.

Hilson, G. 2012. Family hardship and cultural values: Child labor in Malian small-scale gold mining communities. *World Development* 40(8): 1663–1674.

Hilson, G., Ackah-Baidoo, A. 2011. Can microcredit services alleviate hardship in African small-scale mining communities? *World Development* 39(7): 1191–1203.

Hilson, G., Garforth, C.J. 2012. 'Agricultural poverty' and the expansion of artisanal mining: Case studies from West Africa. *Population Research and Policy Review* 31(3): 435–464.

Hilson, G., Hilson, A., Maconachie, R. 2018. Opportunity or necessity? Conceptualizing entrepreneurship at African small-scale mines. *Technological Forecasting and Social Change* 131: 286–302.

Hilson, G., Laing, T. 2017. Gold mining, indigenous land claims and conflict in Guyana's hinterland. *The Journal of Rural Studies* 50: 172–187.

Hilson, G., Maconachie, R. 2020. Artisanal and small-scale mining and the sustainable development goals: Opportunities and new directions for sub-Saharan Africa. *Geoforum* (in press). doi:10.1016/j.geoforum.2019.09.006.

Hilson, G., McQuilken, J., Perks, R. 2019. *State of the artisanal and small-scale mining sector.* DELVE. World Bank, Washington, DC.

Hobbs, J.C. 2005. Enhancing the contribution of mining to sustainable development. In: *Sustainable minerals operations in the developing world* (eds. B.R. MArker et al.), pp. 9–24, Geological Society, London, Special Publications, 250.

Hollaway, J. 1991. *Role of small-scale mining in Africa: Building the informal sector, in alluvial mining.* Institution of Mining and Metallurgy, Elsevier Science, London.

IMPACT. 2019. *AFECCOR: Supporting artisanal gold mining communities to access savings and credit.* IMPACT, Ottawa.

International Labour Organisation (ILO). 1999. *Social and labour issues in small-scale mines.* Report for Discussion at the Tripartite Meeting on Social and Labour Issues in Small-Scale Mines. International Labour Organization, Sectoral Activities Program, International Labour Office, Geneva.

International Labour Organization (ILO). 2001. *Safety and health in small-scale surface mines: A handbook.* International Labour Organization, Geneva.

Jenkins, P.C. 2016. *The politics of mine closure and sustainable development at the Ok Tedi Mine, Papua New Guinea.* Thesis submitted for the degree of Doctor of Philosophy. Australian National University, Canberra.

Jennings, N.S. 2003. Addressing labour and social issues in small-scale mining. In: *The socio-economic impacts of artisanal and small-scale mining in developing countries* (ed. G.M. Hilson), p. 151–160. A.A. Balkema, The Netherlands. Retrieved from https://libro.eb20.net/Reader/rdr.aspx?b=209615438.

Labonne, B. 1994. Small- and medium-scale mining: The Harare seminar and guidelines. *Natural Resources Forum* 18(1): 13–16.

Moore, S. 2018. *Ethiopian extractive industries transparency initiative (EEITI): Third EEITI report.* Extractive Industries Transparency Initiative, Oslo.

Muff, R., Otgonbaatar, A., Bolormaa, B., Batsukh, E. 2016. *Annual bulletin of the geology and mining sector of Mongolia.* Mineral Resources Authority of Mongolia, Ulaanbaatar.

O'Faircheallaigh, C., Corbett, T. 2016. Understanding and improving policy and regulatory responses to artisanal and small scale mining. *The Extractive Industries and Society* 3(4): 961–971.

Organization for Economic Cooperation and Development (OECD). 2013. *OECD due diligence guidance for responsible supply chains of minerals from conflict-affected and high-risk areas.* Organization for Economic Cooperation and Development (OECD), Paris.

Pasha, S., Wenner, M.D., Clarke, D. 2017. *Toward the greening of the gold mining sector of Guyana.* Inter-American Development Bank, Washington, DC.

Responsible Jewellery Council (RJC). 2013. *Responsible jewellery council standards guidance.* Responsible Jewellery Council, London.

Schoneveld, G., Chacha, M., Njau, M., Jønsson, J., Cerutti, P.O., Weng, X. 2018. *The new face of informality in the Tanzanian mineral economy: Transforming artisanal mining through foreign investment?* IIED Research Report, London.

Seccatore, J., de Theije, M. 2017. Socio-technical study of small-scale gold mining in Suriname. *Journal of Cleaner Production* 144: 107–119.

United Nations. 2016. *Mapping mining to the sustainable development goals: An atlas.* United Nations, New York.

United Nations Environment Program (UNEP). 2012. *Analysis of formalization approaches in the artisanal and small-scale gold mining sector based on experiences in Ecuador, Mongolia, Peru, Tanzania and Uganda.* United Nations Environment Program, Geneva.

Vélez-Torres, I., Vanegas, D.C., McLamore, E.S., Hurtado, D. 2018. Mercury pollution and artisanal gold mining in alto Cauca, Colombia: Woman's perception of health and environmental impacts. *The Journal of Environment and Development* 27(4): 415–444.

World Bank. 2005. *The millennium development goals and small-scale mining.* World Bank, Washington DC.

# GOAL 9 Build Resilient Infrastructure, Promote Inclusive and Sustainable Industrialization and Foster Innovation

*John Steen*

## CONTENTS



Sustainable industrialization depends on mined products, but innovation will be needed for mining to be sustainable.

## SUMMARY TABLE: THE COMPLEX RELATIONSHIP OF MINING WITH SDG 9

| Key Stakeholders | Direct Mining Contributions | Contribution to SDG | Key Regulations and public policies | Corporate Policies and Private Programmes |
|---|---|---|---|---|
| Government<br>Local mining communities<br>Automotive industry<br>Energy industry<br>Construction industry | Mined materials for the power and transport infrastructure of the decarbonized global economy<br>Mining development as a vehicle for economic development and construction of legacy infrastructure | Provides the raw materials needed for sustainable industrialization, especially electrification<br>Innovation will enable the expansion of mining to meet these needs<br>Human-centred design in mining technology innovation will enable higher-paying jobs to be retained in mining communities | 'End of pipe' approach to regulation<br>Carbon price signals<br>Geological and environmental data transparency (open data) | Mine to consumer carbon accounting (scope 3 carbon emissions)<br>Shared value strategy<br>Innovation in partnership with local mining supply chains |

## INTRODUCTION

SDG 9 is 'Build resilient infrastructure, promote inclusive and sustainable industrialization and foster innovation.' While innovation comes last in this description, one can argue that innovation will underpin the entire agenda of SDG 9. Without innovation we are left with the current infrastructure technologies and industrialisation models that continue on an unsustainable path.

Mining has a vital relationship with SDG 9. The new infrastructure that will be needed for sustainable energy, transport, accommodation and food systems will require an enormous quantity of mined materials. If the mining industry cannot deliver these inputs at scale and low-cost, then the transition to a sustainable global economy will be jeopardized (Sovacool et al. 2020). Policymakers will need to reconcile the needs of the sustainable economy to obtain critical inputs from an extractive industry with the unavoidable choices in environmental trade-offs. Innovation may diminish the environmental impact of mining, but this impact will never be zero.

Mining can also be seen as a direct catalyst for SDG 9. Mining operations are frequently located in developing economies where they become major employers with an influence over local development and industrialization. In the past, mining has been associated with the 'resource curse' where the economic benefits of mining crowd out broader industrial growth, so that when the mining operation is finished, the economy retreats to a worse position than when the mine started (Sachs and Warner, 1999). Increasingly mining companies are aware of this problem and see themselves as contributors to sustainable economic development through contributions such as training local workers, encouraging enterprise and multi-use infrastructure.

This chapter starts with a discussion of the role of mining in economic development and then proceeds to consider the sectors of the mining industry that will be important for SDG 9 due to their role in renewable energy generation and electric transport. Given the unprecedented demand for mined materials in the twenty-first century, the chapter considers how circular economy principles and inclusive economic development could transform mining to minimize negative environmental and social impacts.

## HOW IS MINING CONNECTED TO DEVELOPMENT CHALLENGES SURROUNDING SDG 9?

Mining provides the raw materials for infrastructure and industrialization. The attainment of a sustainable industrialized economic development path is intrinsically related to the future of mining. Throughout history, materials from the earth have been essential for economic growth and the rise in productivity and living standards that accompany it. Coal and steel fuelled much of the nineteenth- century industrial revolution that transformed societies with new technologies. By today's pace of change in technology, the nineteenth century looks sedate. The enabling technologies that drove the industrial revolution were the steam engine and mass transport, especially rail (Perez, 2002). This required a massive expansion of the mining industry to provide the materials needed for the deployment of these innovations at scale.

Electric power provided a platform for a new wave of innovations from the end of the nineteenth century and into the twentieth century. Along with steel and coal, this required the development of new approaches to mining copper and by the 1920s, large-scale copper mining practices had been developed that could deliver low-price copper to sustain the construction of electricity networks. Continued economic growth through the twentieth century has been dependent on a variety of metals and materials, ranging from coal, oil and gas for energy, to iron, nickel, copper and aluminium for machines and infrastructure. Computers require silicon, gold, palladium and a variety of other elements. The digital economy is thus based upon mined products in the same way that the industrial revolution was. One estimate of the long-term growth trend in copper indicates that this trend is exponential, with consumption doubling every 25 years (Meinert, Robinson and Nassar, 2016).

A consequence of these waves of economic development is the increased consumption of energy, and the production of carbon dioxide as a result. Climate change from $CO_2$ production poses the biggest threat to sustainable industrialization and infrastructure and meeting SDG 9 will require a decoupling of energy consumption from carbon dioxide production on a global scale, in a relatively short period of time. This challenge is recognized explicitly in one of the indicators that is part of SDG 9. Reducing $CO_2$ per unit of value added (SDG 9.4.1) is an important measure of sustainable industrial growth. For example, the $CO_2$ intensity of steel (emission per tonne) has been steady from 2000 to 2019 but needs to decline by 1.9% annually to be in line with the Paris Agreement on climate change

The UN Paris Agreement* target of limiting average global temperature increases to two degrees above preindustrial levels will be difficult to achieve but paradoxically will require an expansion of the mining industry on an unprecedented scale to build renewable energy infrastructure and electric transport (Sovacool et al. 2020). While recycling of mined materials will help, the sustainable twenty-first-century economy will be highly dependent on mined products.

## Renewable Energy and Mined Materials: A Paradoxical Link

Decarbonizing the global economy is an enormous task because it will require replacement of existing energy sources that are based on fossil fuels. The replacement of pre-existing energy sources has never happened in the history of industrialization. Industrialization has historically been a process of the new fuel accounting for capacity additions, rather than replacement of legacy fuels (Columbia Center on Global Energy Policy, 2019). For example, the introduction of coal as a dominant energy source during the industrial revolution did not displace wood-fired energy, and oil in the twentieth century did not displace coal. To achieve the needed cuts in $CO_2$ emissions, renewable energy accounts for new capacity in a growing economy but it must also replace legacy fossil fuel energy sources as well. Existing energy infrastructure will need to be replaced, in conjunction with the addition of new energy infrastructure.

To understand how much metal this will require on the energy generation side to both add to energy capacity and replace existing fossil fuel capacity, we can use estimates of metal usage for wind and solar power. Using a figure of 3,000 tonnes of copper per GW for wind power and 4,000 t/GW for solar indicates that global installation of new wind power between 2019 and 2024 will require 390,000 tonnes of copper, and 2,500,000 tonnes of copper for solar PVs (Wood Mackenzie, 2019; Wood Mackenzie, 2020). Hence, based on these forecasts, to meet the needs of the solar and wind industries alone, copper mining will need to add another 3% of production capacity per year.

Steel is an important material for twenty-first-century industrial infrastructure and renewable energy infrastructure also requires steel, which is a product of iron ore and metallurgical coal. A 3MW wind turbine uses around 300 tonnes of steel (Willburn, 2011), with the complication of $CO_2$ production as part of the steel-making process where carbon from coking coal is used to remove oxygen from the iron ore. In the future it may be possible to remove this oxygen with hydrogen, with water being a harmless by-product, but in the short to medium term the link between $CO_2$ emissions and steel production will remain (Åhman et al. 2018).

* The Paris Agreement is an agreement within the United Nations Framework Convention on Climate Change (UNFCCC), dealing with greenhouse-gas-emissions mitigation, adaptation and finance, signed in 2016. As of November 2019, 195 UNFCCC members have signed the agreement, and 187 have become party to it.

Another interesting example to illustrate these paradoxical links is the case of neodymium, electric motors and generators. This rare-earth metal is used to improve the performance of magnets and will be needed for turbines and electric vehicle motors. A typical 3MW wind turbine contains 120kg of neodymium (Willburn, 2011). Like other 'rare earths' the occurrence of neodymium is not geologically rare. It occurs, however, in low concentrations and is difficult to extract from ores (Habashi, 2013). Radioactive thorium is a hazardous by-product of neodymium production and nearly all rare earths co-occur with uranium and thorium (Gupta and Krishnamurthy, 1992; Jordens, Cheng and Waters, 2013). Most neodymium refining and magnet production occur in China where the industry has been allowed to pollute large areas of land (Weng, Jowitt, Mudd and Haque, 2015; Huang et al. 2016). Innovation in rare-earth production, including neodymium, will be less focused on mining and more on processing. If new processing methods are discovered then neodymium production can become common outside of China where geopolitical tensions will not pose supply threats. Supply chain security for a wide range of high-tech applications, including electric motors and generators, will be thus easier to achieve. This must be a high-priority R&D target for governments.

For electrification to succeed, there will also need to be storage of energy – and batteries will play a large part in making renewable energy systems reliable. Battery technology is steadily evolving but has the main requirements of lithium, nickel and cobalt. Lithium is in plentiful supply in different forms around the world, but nickel is less common, and around 60% of the world's cobalt is sourced from the Democratic Republic of Congo. In the DRC, small-scale cobalt mining has presented problems for worker health and safety, and the trade in cobalt has funded armed conflict between political rivals (Nkulu et al. 2018). At the present time over 85% of cobalt used for battery production is controlled by Chinese mining companies (Vella, 2018). The economic implications of critical metals being in the control of geopolitical rivals have only just been realized by Western governments and broad cooperative alliances between the USA and mining-intensive countries such as Canada and Australia are now under development (U.S. Mission to Canada, 2019)

## SUSTAINABLY EXPANDING THE MINING INDUSTRY IN THE TWENTY-FIRST CENTURY

Circular economy principles provide an overall framework for the sustainable expansion of the mining industry to meet growing demands (Webster, 2017). These principles include: designing out waste and pollution, keeping products and materials in use, and regenerating natural systems. The circular economy approach is a challenge to traditional concepts of mining, which are based on linear business models of production and consumption in which profit from the mine is maximized by increased consumption of materials. Mines are increasingly adopting renewable energy due to the rapidly declining cost of electricity from these sources, but the products of mining will need to become renewable as well.

To keep materials in use within circular industrial systems, mining companies, and the mining industry as a whole, will need to have oversight and responsibility for the use of their products. Already we are seeing moves in this direction with the experimental use of blockchain technology for tracking the provenance and use of metals, especially for electric vehicles (Mining Review Africa, 2019). The recent decisions by BHP to be held accountable for the $CO_2$ emissions of its customers (scope 3 emissions) are also significant and are part of the transition to circular economy principles in the mining sector (BHP, 2018). Technologies for tracking materials and their environmental footprints through the supply chain will be an important enabler of the circular economy in mining.

One way that the mining industry can design out waste is through precision mining. As ore grades fall, more waste rock is produced under conventional mining methods to extract the desired materials. This will be a significant problem for all mining activities but especially for those that produce wet tailings as these will need to become larger, with potential environmental damage through spills and wall collapse. Precision mining involves the use of advanced mining technology to target the areas of rock that are valuable and leave the remainder in the ground. This is made possible by automated mining equipment that can work in narrow tunnels without the need for mine ventilation. Such techniques are guided by advanced geological and ore body modelling, where breakthroughs are being assisted by artificial intelligence to make better use of drilling data. When miners explore deeper ore-bodies to meet market demand, the use of robotics will become more common to reduce human exposure to dangerous working conditions.

Mining companies are beginning to use innovative approaches to regenerating natural systems, using science to reconstruct ecosystems. For example, Canadian mining company Teck has reclaimed an old tailings pond at Highland Valley Copper, British Columbia, and turned it into a fishing lake. Once the phosphorous and nitrogen nutrients in the lake were rebalanced, bacteria were added, then algae and zooplankton, and finally fish (Teck, 2016). This life-of-mine approach to environmental management will be important for implementing the circular economy.

## MUCKAHI UNDERGROUND MINING SYSTEM: MINE INNOVATION FOR ECONOMIC INCLUSION

As underground mines become deeper and ore grades fall, there is an urgency to increase productivity in underground mining through innovation (Steen, Macaulay, Kunz and Jackson, 2018). The dominant thinking in the industry is that autonomous vehicles, enabled by underground internet and electric power, will mine the underground mine of the future. Tunnels can be narrower, operations will be continuous rather than in shifts and workers can be taken out of mines. While this future will enable the expansion of ore production from deeper ore bodies at lower costs, it has the drawback of removing jobs for workers. While there is an argument that the overall number of jobs in the mining sector will increase as technology enables industry expansion (EY, 2018), these are not the entry-level jobs that workers in developing economies rely upon for wages.

However, there may be good reasons to design mining innovations to keep workers engaged in mining operations when the opportunity to train workers to perform highly skilled technical jobs is limited. The first of these is that local employment is an important part of a mine's social licence to operate. In rural economies the primary employer is mainly small-scale agriculture and mining offers a chance for a higher wage, which is often spent on local businesses and results in economic development around the mine (Aragon and Rud, 2013).

Another reason to engage local employees in mine operations is that the supply chain expertise and infrastructure to support high-tech mines may not be available in remote locations. A fully automated mine might be more productive than a mine that employs local workers, but the ability to improve performance and minimize disruption could be higher in the labour-intensive operation, which would also have lower capital costs.

Ideally, mining innovation for development improves mine productivity, thus enabling higher wages for workers, but also keeps local employees in the mine. This is an innovation design challenge that is rarely seen in the mining industry; however, it is possible. An example of innovation that generates inclusive productivity growth in an underground mine is the Muckahi underground excavation system that has been developed for gold mining in Mexico by Torex Gold Resources (Moore, 2019). The conventional thinking around underground mining is that tunnels need to be wide enough to allow vehicles to pass each other so that after a blast, the tunnel can be 'mucked out' with the removal of rock.

These wide tunnels raise the cost of underground mining, but Torex decided to take a lateral approach to the problem by designing a dual-carriage overhead monorail that could move rock from the blast site quickly but be simple enough to use so as to keep local workers on the mine. It carries the additional benefits of allowing the tunnels to be steeper, thereby accessing the required depth faster. Electrification reduces the need for the ventilation that removes the heat and particles associated with diesel equipment, which also reduces the capital and operating costs of the mine.

## CONCLUSIONS

Innovation, upgrade infrastructure and retrofit industries to make them sustainable, with increased resource-use efficiency and greater adoption of clean and environmentally sound technologies and industrial processes (As SDG 9 and Target 9.4 indicates), must be at the core of economic progress. All of the infrastructure and economic development that we enjoy today is a result of some form of innovation. This will also be true for restructuring the economy to be sustainable and reach zero net carbon emissions in order to stabilize the climate and restrict warming to a level that minimizes disruption to the international community. Innovation will be needed in mining and mineral processing to achieve this goal but other forms of innovation that are overlooked will be just as important. Public policy innovation will be necessary to design the regulatory regimes that send price signals for pollution through the mining value chain. An example of how this might occur is the use of sensors to track $CO_2$ production and the information captured by these sensors is entered into a blockchain so that regulators and consumers can understand the environmental impact of the product that is being produced. The feedback from consumers and price-signals from regulators will accelerate the adaptation of the mining industry to a sustainable twenty-first-century economy.

## CASE ANALYSIS

## BUILD RESILIENT INFRASTRUCTURE, PROMOTE INCLUSIVE AND SUSTAINABLE INDUSTRIALIZATION AND FOSTER INNOVATION

*Democratic Republic of the Congo (DRC)*

**Ben Radley**

### Structural Limitations of Mining-Led Industrialization

Over the last decade, multinational mining corporations have been slowly returning to the eastern parts of the Democratic Republic of the Congo (DRC) that were the worst affected by the Congo Wars (1996–2002). In 2012, the Canadian gold miner Banro was the first to restart industrial production in the region following the sector's collapse in the 1990s, through its flagship Twangiza mine. Located in South Kivu Province, the mine produces around 3,500 kilograms of gold per year and employs around 1,300 people.

Recent mining industry restructuring away from vertical integration and towards outsourcing the supply of goods and services to independent firms has fuelled fresh hopes that in low-income country contexts like the DRC, the industry now carries the potential to drive industrialization and structural transformation. Yet while Banro has indeed been outsourcing a range of activities and services at Twangiza to independent firms, two structural constraints limit the mine's ability to promote inclusive or sustainable industrialization in the DRC.

First, the machinery required by industrial gold mining today is relatively narrow and highly specialized. Of the 52 million USD held at Twangiza in fixed assets, around 31 million USD represented the processing plant and heavy mobile machinery (such as dump trucks and excavators). The plant – comprising a three-stage crushing system, two ball mills, a carbon-in-pulp chemical treatment facility, a gold storage room and a smelting laboratory – was purchased second-hand and transported by sea from Australia. It was then delivered by road through the Kenyan port of Mombasa, via 140 40-foot containers, for reassembly by engineers on-site at Twangiza. Since reassembly, the plant has had little need for additional inputs that could meaningfully stimulate the Congolese economy.

Second, the DRC is wholly dependent upon the external manufacture of the machinery in use at Twangiza, the majority of which continues to come from Northern America, Europe, Japan, Australia and South Africa. Both during and after mine construction, Banro made exclusive use of overseas firms and suppliers. Of the 37 manufacturers listed on Twangiza's asset inventory, 32 were American, European or Japanese. With the exception of South Africa, no Congolese or other African firms appeared on the list. In the area of subcontracting, only 2 of the 12 firms subcontracted by Banro at Twangiza were Congolese, and both of these were in low value-added activities.

The Twangiza mine is, then, strongly disconnected from the surrounding Congolese economy. As the provincial representative of the Congo Business Federation said, 'Banro hasn't had much of an impact on the private sector here. Banro could close tomorrow and we wouldn't know.' The thinness and specialized nature of the required capital to keep Twangiza operational and the mine's dependency on externally manufactured technology and firms, places a set of structural constraints on the sector's ability to contribute to broader-based and sustainable processes of industrialization, particularly in low-income country contexts like the DRC.

## REFERENCES

Åhman, M., Olsson, O., Vogl, V., Nyqvist, B., Maltais, A., Nilsson, L.J., Hallding, K., Skånberg, K. and Nilsson, M. 2018. *Hydrogen Steelmaking for a Low-Carbon Economy: A Joint LU-SEI Working Paper for the HYBRIT Project.* EESS report no 109 and SEI working paper WP 2018–2007. Stockholm Environment Institute: Stockholm, and Lund University: Lund.

Aragon, F. and Rud, J.P. 2013. Natural resources, and local communities: Evidence from a Peruvian gold mine. *American Economic Journal: Economic Policy* 5(2): 1–25.

BHP. 2018. *Scope 3 Emissions Calculation Methodology.* BHP: Melbourne.

Columbia Center on Global Energy Policy. 2019. *Energy Transition Fact Sheet.* www.energypolicy.columbia.edu/sites/default/files/file-uploads/Energy%20Transition%20Fact%20Sheet.pdf. Accessed 1/1/2020.

EY. 2018. *The Future of Work: The Changing Skills Landscape for Miners.* EY: Sydney.

Gupta, C. and Krishnamurthy, N. 1992. Extractive metallurgy of rare earths. *International Materials Reviews* 37(1): 197–248.

Habashi, F. 2013. Extractive metallurgy of rare earths. *Canadian Metallurgical Quarterly* 52(3): 224–233.

Huang, Z., Zhang, G., Pan, A., Chen, F. and Zheng, C. 2016. Protecting the environment and public health from rare earth mining. *Earth's Future* 4(11): 532–535.

IEA. 2019. *Tracking Industry.* IEA: Paris. https://www.iea.org/reports/tracking-industry-2019.

Meinert, L., Robinson, G. and Nasar, N. 2016. Mineral resources, reserves, peak production and the future. *Resources* 5(1): 14.

Mining Review Africa. 2019. *How Blockchain Is Shaping Ethics and Morals in Mining.* www.miningreview.com/battery-metals/blockchain-ethics-morals-mining/. Accessed 1/1/2020/.

Moore, K. 2019. *The Dual Overhead Monorail Based Muckahi Mining System.* Mining World. www.miningworld.com/the-dual-overhead-monorail-based-muckahi-mining-system/. Accessed 1/1/2020.

Nkulu, B., Casa, L., Haufroid, V. et al. 2018. Sustainability of artisanal mining of cobalt in DR Congo. *Nature Sustainability* 1(9): 495–504.

Perez, C. 2002. *Technological Revolutions, and Financial Capital: The Dynamics of Bubbles and Golden Ages.* Edward Elgar: Cheltenham.

Sachs, J. and Warner, A. 1999. The big rush, natural resource booms and growth. *Journal of Development Economics* 59(1): 43–76.

Sovacool, B.K., Ali, S.H., Bazilian, M., Radley, B., Nemery, B., Okatz, J., and Mulvaney, D. 2020. Sustainable minerals and metals for a low-carbon future. *Science* 367(6473): 30–33.

Steen, J., Macaulay, S., Kunz, N. and Jackson, J. 2018. Understanding the innovation ecosystem in mining and what the digital revolution means for it. In: *Extracting Innovations: Mining, Energy, and Technological Change in the Digital Age.* Clifford, M., Perrons, R., Ali, S., and Grice, T. (Eds). CRC Press: New York.

U.S. Mission to Canada. 2019. *First Meeting of the U.S.- Canada Critical Minerals Working Group.* ca.usembassy.gov/first-meeting-of-the-u-s-canada-critical-minerals-working-group/. Accessed 1/1/2020.

Teck. 2016. *Reclamation Example: From a Tailings Pond to a Trout Pond.* https://www.teck.com/news/stories/2016/reclamation-example---from-a-tailings-pond-to-a-trout-pond. Accessed 2/13/2020.

Vella, H. 2018. Inside China's moce to monopolise cobalt. *Mining Technology.* www.mining-technology.com/features/inside-chinas-move-monopolise-cobalt/. Accessed 1/1/2020.

Webster, K. 2017. *The Circular Economy: A Wealth of Flows.* 2nd edn. Elen Macarthur Foundation: Cowes, U.K.

Weng, Z., Jowitt, G., Mudd, G. and Haque, N. 2015. A detailed assessment of global rare earth element resources: Opportunities and challenges. *Economic Geology* 110(8): 1925–1932.

Wilburn, D.R. 2011. *Wind Energy in the United States and Materials Required for the Land-Based Wind Turbine Industry from 2010 through 2030.* U.S. Geological Survey Scientific Investigations Report 2011–5036.

Wood Mackenzie. 2019. *Global Wind Turbine Technology Trends 2019.* Wood Mackenzie: London.

Wood Mackenzie. 2020. *Global Solar PV Market Outlook Update: Q1 2020.* Wood Mackenzie: London.

# GOAL 10 Reduce Inequality within and among Countries *The Case of Antofagasta, the Main Mining Region of Chile*

*Cristian Parra and Saleem Ali*

## CONTENTS



During the last century, the global extractive-mining industry has been consciously working to improve the understanding of its socio-economic effects and how the opportunities of the sector are transformed into sustainable development and social progress. It is possible to observe significant progresses in terms of: regulations, policies and standards; the engagement processes among companies, governments and affected communities; the due diligence to obtain permits and approvals; the designing of social management plans and mitigation and offsetting measures; and how the sector is dealing with core concerns such as Human Rights, indigenous participations, bribery and corruption, accountability principles and transparency among others. Nevertheless, despite how progress in many areas is moving in the right direction, the expectations and demands from civil society are also growing, and one of the key remaining concerns is the understanding of the final effects of the mining industry on personal income distribution or income inequality.

Recent research by Davis (2019) suggests that in comparison with the oil industry, mining economies were found to cause higher national income inequality on a gross, pre-tax basis and a net, post-tax and post-redistribution basis. Thus the obligation of the mining industry to consider social mobility and wealth distribution prospects of their operations in the context of SDG 10 becomes even more salient. This chapter considers the case of Chile, a country that faced massive political turmoil in 2019, which was frequently blamed on inequality despite its mineral fortunes.

Hence, in the context of the SDG 10 and the historical and strong connections between the industry and Chilean society, and the global reputation of Chile as a top jurisdiction for mining investment (Fraser's Investment Attractiveness Index for Minerals 2018), this chapter explores the relationship between Chilean mining activity and the evolution of income inequality and reflects on the necessary next steps to assess income distribution beyond the traditional indicators. The analysis is presented for the period 1990 to 2018 and is based on official statistics from Antofagasta Region, which is the most important mining region of Chile and Latin America.

## INTRODUCTION

Chile is the principal supplier of copper worldwide, with 5.4 million metric tonnes per year, equivalent to 34% of world copper production (COCHILCO 2018). Chilean copper reserves account for 25% of worldwide reserves (USGS 2018). The most important mining region in Chile is Antofagasta, which represents 75% of the Chilean mining GDP and 75% of Chilean copper production (Banco Central de Chile 2018). During the last three decades, the mining activity in Antofagasta grew in terms of production, foreign investment and new large-scale mining projects (Consejo Minero de Chile 2018).

To date, the mining assets invested account for more than US$30 billion; in the last two decades foreign investment materialized was over US$15 billion; and the contribution to state income topped 25% (considering direct and indirect effects). Resource development is undertaken by a mix of both Chilean state-owned companies such as CODELCO, and major global resource corporations including BHP Billiton, Anglo-American, Glencore, Freeport-McMoRan and Barrick Gold among others.

In this context, Chile and Antofagasta are extremely interesting cases for analysis to understand the potential effect of mining activities on income inequality and SDG 10. Chile has been highlighted as one of the best examples in terms of positive socio-economic progress in Latin America and as the only country that has already accomplished most of the previous Millennium Development Goals and the target of halving extreme poverty (PNUD 2015). A comparative analysis with other non-mining regions of Chile, shows a relative better socio-economic performance of Antofagasta (Parra 2012).

## RESULTS

The analysis of the personal income distribution of Antofagasta using the ratio 20:20 (ratio of the personal income share held by the highest 20% compared to the lowest 20%) has improved significantly over the last 30 years (Target 10.1: Reduce income inequalities). This performance greatly exceeds the progress of Latin America and Chile during the same period, and places Antofagasta at a similar level of income distribution of Spain or Australia.

**Personal Income Distribution in Mining Regions of Chile (Ratio 20:20)**
**Ratio of the personal income share held by the highest 20% to the lowest 20%**

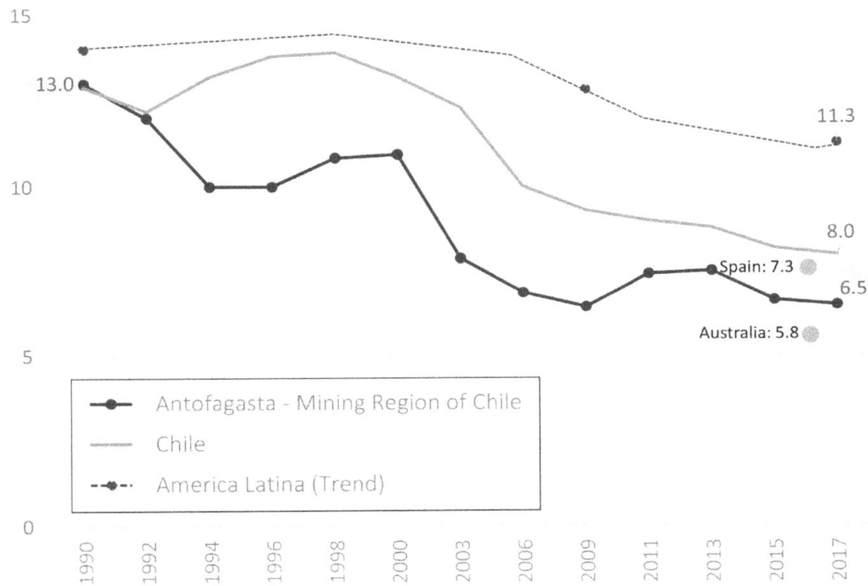


Considering the social public policies in Chile are the same across the country (Chile is not a federal state) for education, health and direct social support, it is possible to understand this positive trend as a direct result of two main factors: mining activity and a stable and non-discriminatory framework of mining policies. Hence, behind these positive trends in Antofagasta, the explanation should consider at least three areas related to mining's economic growth: 1) economic linkages, 2) re-investment of net profits in new projects and sustainable capital and 3) direct and indirect effect on employment.

## BEYOND THE ANALYSIS OF INCOME DISTRIBUTION

In the context of Agenda 2030 and SDG 10, and despite the positive relationship between mining activities and income distribution improvement, it is important to consider an analysis beyond the specific indicators.

### What Is Fair in Terms of Income Distribution?

Income inequality improvement in Antofagasta has been a result of pure economic growth, however it is difficult to find evidence related to coordinated efforts among policymakers, interested groups, civil society and mining companies to develop specific policies, strategies and actions focused on specific social aspects to transform mining opportunities into better personal income conditions. Hence, the relevant question is: Should the income distribution of Antofagasta be considered as a reasonable distribution, considering the tremendous economic growth of the last years? The answer is probably no. Because of the significant economic growth of Antofagasta, the final outcome in terms of inequality could have had a better result.

## A Holistic Approach to Evaluate Personal Income Distribution

Evaluations of personal income distribution can evidence positive trends on basic parameters such as ratio 20:20; however, this is not enough. The analysis must be extended to inequality and non-discriminatory conditions and include a more holistic view of different factors such as education, work conditions and opportunities, non-discriminatory employment policies and support for the most vulnerable people, and evaluate if the economic opportunities are offering real access to a decent standard of living.

Different mining regions can show socio-economic growth, but in the context of SDG 10, the real challenge is to address the complexity of sustainable development challenges.

## Considering the SDG 10 Targets in the Context of the Antofagasta Chile Case

It is important to note that economic growth based on massive mining activities has the capacity to positively impact different socio-economic indicators; however, these activities do not have the capacity to reduce to zero or eradicate poverty. Thus, considering poverty still affects around 7% to 10% of the population in Antofagasta, assessing personal income distribution (despite the positive trend) requires a balanced evaluation.

If we are to now qualitatively evaluate the situation in the mining region with respect to the key targets for SDG 10, we arrive at the evaluation shown in Table 10.1.

## Final Reflections in the Context of Resource-Rich Countries

Even with the recognition of personal income distribution improvement and highly promoted socio-economic benefits of the industry of non-renewable natural resources (in terms of GDP, investment, exports, local procurement, employment, technological knowledge, direct positive community impacts and fiscal incomes), it is easy to observe a high level of socio-environmental conflicts and growing difficulties to maintain the social licence to operate.

The social unrest and turbulence of Chile initiated in October 2019 on behalf of inequality and social conditions, with the support of mining unions and labour force groups, is a clear signal to definitely look beyond the apparent economic success of a mining region. It is time to work for a more effective set of policies and programs to transform the extractive industry opportunities into real and inclusive

**TABLE 10.1**
**SDG 10 Relevant Targets – Evaluation in Antofagasta, Chile**

| SDG 10 Target | Evaluation in Antofagasta, Chile |
|---|---|
| By 2030, progressively achieve and sustain income growth of the bottom 40% of the population at a rate higher than the national average. | There is a high correlation between mining income growth of the bottom 40%, however as there are just a few new mining projects and expansion, it is not possible to anticipate additional improvements. |
| By 2030, empower and promote the social, economic and political inclusion of all, irrespective of age, sex, disability, race, ethnicity, origin, religion or economic or other status. | There are different corporate efforts to empower people and promote social inclusion. The industry must continue to working in that direction. |
| Ensure equal opportunity and reduce inequalities of outcome, including by eliminating discriminatory laws, policies and practices and promoting appropriate legislation, policies and action in this regard. | There are different policies oriented to reduce inequalities, however it will be absolutely necessary to work on education and promoting new values and views. |
| Adopt policies, especially fiscal, wage and social protection policies, and progressively achieve greater equality. | Decent salaries and social protection of employees has been on the negotiation table for a long time. It is time to move to a more open discussion. |

growth, socio-economic progress and to promote equality from a more holistic perspective (SDG 10, Target 10.4: Adopt fiscal and social policies that promotes equality, and, using indicators such as 10.4.1: Labour share of GDP, comprising wages and social protection).

Latin America and the Caribbean, with a heavy dependence on exports of mining and natural resources, continue to be the most unequal. More than 187 million people continued living in poverty and 62 million in extreme poverty in 2017 (ECLAC 2018). To overcome this will necessarily require a new generation of fiscal policies for taking action in the social arena and ensuring that no one is left behind, precisely as indicated by the 2030 Agenda for Sustainable Development and SDG 10; it will be necessary to understand that equality plays a critical role as the driver of sustainable development, contributing to innovation, increased productivity and environmental sustainability for future generations.

## REFERENCES

Banco Central de Chile (2018). *Compendio de estadísticas económicas*. Santiago, Chile.

CASEN (2018). Ministerio de planificación de Chile. *Resultados encuesta CASEN.*

Consejo Minero de Chile (2015). Reporte anual. Santiago, Chile.

Davis, Graham A. (2019). Large-sample evidence of income inequality in resource-rich nations. *Mineral Economics*. doi:10.1007/s13563-019-00207-1.

ECLAC (2018). *Statistical Yearbook for Latin America and the Caribbean*. United Nations Publications, Santiago, Chile.

Gobierno de Chile/PNUD (2015). Tercer Informe sobre los Objetivos de Desarrollo del Milenio.
Parra, C. (2008). *Desarrollo Humano y Socioeconómico en la Región de Tarapacá 1990–2007.* SCM Collahuasi, Tarapacá, Chile.
USGS (2018). Mineral commodity survey.

## BIBLIOGRAPHY

AIA (2014). *Asociación de Industriales de Antofagasta.* www.aia.cl.
COCHILCO (2019). Reporte anual. Santiago, Chile.
Gobierno de Chile (2016). *Descripción/Características de la Región.* http://www.goreantofagasta.cl/index.php/historia-region.
Lagos, G. and E. Blanco (2010). Mining and development in the region of Antofagasta. *Resources Policy* 35(4): 265–275.
Larde, J., E. Chaparro, et al. (2008). *El aporte del sector minero al desarrollo humano, El Caso de la Región de Antofagasta.* División de Recursos Naturales de la CEPAL, UN. Santiago, Chile.
Parra, C. (2009). *El caso de la Provincia de San Juan y Mina Veladero.* San Juan, Argentina.
Parra, C. and D. Franks (2011). *Monitoring Social Progress in Mining Zones - The Case of Antofagasta and Tarapacá, Chile.* Barrick Gold, SRMining Santiago, Chile.

GOAL 11

# Make Cities and Human Settlements Inclusive, Safe, Resilient and Sustainable

*Laura García Jaramillo*

## CONTENTS



## IMPACTS OF MINING ON SDG 11

### Natural Synergies and Positive Impacts

Mining indirectly contributes to making cities and settlements inclusive, safe, resilient and sustainable by producing many of the raw materials used in construction and for the energy transition to cleaner power generation; by creating jobs and contributing to creating an enabling environment for businesses; and by paying taxes and royalties that enable governments to fund social development programs.

The presence of mining projects and operations revitalises the economy through the mine's value chain, indirectly helping to transform productive capacities; and the infrastructure that is built or improved for mining projects and operations can improve people's access to the benefits of development, and improve connectivity and access to public and social services. The construction of a mining project or a mining operation can also bring more security and contribute to peace-building as there are usually public or private security forces that provide security to the area of influence.

Stakeholder Engagement, Corporate Social Responsibility and Socio-Economic Development programs constitute a direct way in which mining could contribute to SDG 11. When strategically and holistically structured and implemented, these programs could encourage communities to organise and empower themselves to participate in decision-making; social investment programs are usually aimed at meeting communities' most urgent and basic needs (such as infrastructure, access to public services, health and education services, among others); and socio-economic development programs may contribute to inclusiveness and resilience as they are usually focused on capacity and skill-building that improve access to employment, and are targeted especially on vulnerable groups, contributing to improve their livelihoods and overcome dependence on only one economic activity.

Some of the most tangible positive impacts of mining on SDG 11 that can be drawn from the sustainability reports of the five major companies in 2018*, can be outlined as follows.

## Sustainable Development and Resilience

Resilient† communities have better protection against climate change and environmental shocks; and they work to reduce risks and prevent crises (whether from man-made or natural causes).

The natural synergy between mining and climate change relates to the production of the materials for the energy transition, as mentioned earlier. Furthermore, major mining companies are moving towards carbon-free operations and creating new technologies to operate waterless operations and electric fleets. This represents a benefit for the neighbouring communities that in the future will experience fewer environmental impacts from mining operations.

Very common and well-established practices in mining, such as safety, health and environmental risk and impact management and emergency preparedness and response, could be considered as another natural contribution of mining to sustainable communities, specifically to resilience and safety, and to the prevention of crisis, although it is related exclusively to risks created by the specific operation and poses risks that will be further analysed.

Due to increasing social scrutiny, and to events such as those in Brazil where two tailings dams have collapsed in the past two years, leading to the loss of more than 200 lives and large material losses, mining associations and major mining companies are increasing their efforts to integrate into their risk and impact management systems those that can potentially affect the local communities, and are

* www.mining-technology.com/features/the-top-ten-metals-and-mining-companies/

† "Resilient cities offer a diverse range of employment and livelihood opportunities; more equitable access to resources; better protection against climate change, economic and environmental shocks; peaceful settlement of disputes; and democratic governance. They work to reduce risks, prevent crises (whether from man-made or natural causes), avert major development setbacks and promote human security, while helping address humanitarian, peace-building and longer-term development efforts globally." www.undp.org/content/undp/en/home/librarypage/poverty-reduction/sustainable-urbanization-strategy.html

implementing methodologies and strategies to facilitate community's involvement in the response planning.

According to the International Council on Mining and Metals (ICMM):

> while industry has a responsibility to be prepared for emergencies through internal mechanisms, it's equally important to work with communities living near mine sites to increase their understanding of potential threats to safety and security. These threats can be both actual – from direct and indirect exposure to risks – or psychological – caused by fear of the unknown impacts of a potential incident. Both real and perceived risks damage social confidence and trust in the industry [...] through implementation of a robust emergency preparedness mechanisms, mining and metals companies are able to responsibly respond to potential hazards in a way that is timely and sympathetic to the strengths and vulnerabilities of operations and communities.*

The most robust methodology for the integration and involvement of communities into risk and impact management is the APELL Process (Awareness and Preparedness for Emergencies at Local Level), developed by the United Nations Development Program (UNDP), which

> aims at creating a cohesive and resilient community in the face of technological or natural hazards through raising awareness and agreement on roles and responsibilities of all community stakeholders in potential preparedness and response measures. This is achieved by assisting decision-makers and technical personnel to increase community awareness and to prepare coordinated response plans involving industry, government, and the local community, in the event that unexpected events should endanger life, property or the environment.†

Companies such as Anglo American integrate into their social policy the management of social impacts, social incidents and community health and safety, in line with the IFC Performance Standard 4 Community Health, Safety and Security, in an effort that seeks to identify, assess, prevent and manage risks and impacts on communities with the involvement of the potentially affected stakeholders. The company has implemented the APELL methodology in Chile, in partnership with the UNDP, to improve the community's ability to react to a potential tailings dam failure or burst.

There is an additional component to SDG 11 related to promoting human security, while helping address humanitarian, peace-building and longer-term development efforts globally. Mining has a substantial role to play in this area through the security function and the implementation of the Voluntary Principles on Security and Human Rights (VPSHR). In order to secure their facilities, mining companies usually bring public or private forces. Even though this poses risks on HR (that will be looked at in another subsection of this chapter), on the positive side, implementation of VPSHR

* www.icmm.com/en-gb/health-and-safety/safety/emergency-preparedness-response

† www.unenvironment.org/explore-topics/disasters-conflicts/what-we-do/preparedness-and-response/awareness-and-preparedness#methodology

and the UN Guiding Principles on Business and Human Rights helps to protect the right to peaceful assembly, to freedom of expression, and, when implemented fully, helps the protection of defenders of human rights.

In their sustainability report for 2018, Glencore reports their support of peacebuilding in Colombia, a country that recently signed a peace accord with the oldest guerrilla group in the world that concluded in their demobilisation and transformation into a political party. Glencore expresses it is supportive of the steps taken by the government through: strengthening local capacities; implementing integrated rural development projects on non-mining lands; promoting a corporate culture of peace; and responsibly managing its impacts within the framework of the UN Guiding Principles on Business and Human Rights and the UN Voluntary Principles on Security and Human Rights. This is a very sensitive area that is not free of risk and conflicts, as will be analysed further in this chapter.

Once again, widespread knowledge and acceptance of their rights present a community with a significant opportunity for their own development, as it raises levels of awareness that empower them to organise, to act and to grow in reconciliation.

Through their HR and CSR programs, mining companies also contribute to peace-building, bringing governance to remote areas normally beyond the scope of local governments.

## Inclusiveness*

There are evident contributions of mining to enabling all groups of people to contribute to creating opportunities, to sharing in the benefits of development (access to sustainable livelihoods) and creating productive capacities and employment for all, through capacity- and skill-building programs, through local procurement and employing members of local communities, and through identifying and prioritising vulnerable*† and marginalised groups for this type of program.

For example, Glencore and Rio Tinto‡ have implemented indigenous employment programs in Australia, Mongolia and Canada; while Anglo American has implemented youth work-readiness and local workforce development programs, and training plans for non-mining jobs in developing countries such as Peru and South Africa.

* An inclusive city enables all groups of people to contribute to creating opportunities, to share in the benefits of development (access to sustainable livelihoods, legal housing and affordable basic services) and to participate in decision-making. Following UNDP's human development principles of participation, non-discrimination and accountability, cities create productive capacities and employment for all and provide effective and efficient social safety nets to protect those who cannot work or who earn too little (www.undp.org/content/undp/en/home/librarypage/poverty-reduction/sustainable-urbanization-strategy.html).

† Vulnerability refers to the definition of Systemic Vulnerability, which refers to individuals or groups with a reduced or compromised ability to realise or maintain a sustainable livelihood. It describes a pre-existing condition unrelated to a site's presence or activities.

‡ Rio Tinto, for example, reports in its 2018 Sustainability Report that it has 1,484 full time indigenous employees in Australia; that 93% of the workforce at their You Tolgoi operation in Mongolia are Mongolians and 27% employees of the Diavik mine are indigenous.

Resettlement and land management could be considered another contribution of mining to sustainable livelihoods, legal housing and affordable basic services.* When done in line with international standards, such as the International Financing Corporation (IFC) Performance Standards, the International Council on Mining and Metals (ICMM) recommendations, among other best practices, a Resettlement Action Plan (RAP), a Livelihood Restoration Plan (LRP), and even the negotiation process with the affected community, should represent a holistic improvement in their quality of life, as, according to the standards, at the end of the process they should be in better conditions than in their place of origin, should have sustainable livelihoods, should have legal rights over their houses and land and should not be in a vulnerable condition.

In 2018, the five major mining companies conducted resettlement programs in different geographies, most of them in developing countries (Brazil, Colombia, Mongolia, South Africa). Even though all of them report implementing these resettlements in line with IFC Performance Standard 5 on land acquisition and resettlement, they also report that these processes are not finished, as many families refuse to move; furthermore, research shows that this process poses more challenges and risks than benefits, as will be presented in the next subsection of this chapter.

There are also direct and indirect contributions to equitable land development. As access to land is critical for mining operations, companies tend to partner with local governments in order to improve land management according to its potential, and even make partnerships with local governments to ensure legalisation of land, a critical issue in developing countries and those with customary land ownership.

Mining operations also contribute to management of project-induced immigration; this management can have positive spillover effects on local governments' capacity to respond to this phenomenon, by the expansion of the local capacity to provide basic social and public services, improve safety and security and manage risks that otherwise would have remained in the shadows, such as people trafficking, prostitution, drug and alcohol abuse, among others.

Another important component of inclusiveness is the ability of communities to participate in decision-making following the UNDP's development principle of participation, non-discrimination and accountability. Regarding mining activities, this can be translated into the Socio-Political License to Operate (SPLTO).† The social performance standards of the five major mining companies establish stakeholder engagement as a pillar of their relationship with local communities, and most of them also have complaints and grievance mechanisms to ensure a fair interaction with their workforce and external stakeholders.

---

* "Sustainable cities are able to tackle the connected issues of poverty, inequality and exclusion while transforming productive capacities, avoiding the irreversible depletion of social and natural capital and lowering risks arising from shocks. The resource endowments of the poor and their prospects for employment and livelihoods are improved." www.undp.org/content/undp/en/home/librarypage/poverty-reduction/sustainable-urbanization-strategy.html

† For the purpose of this chapter, SPLTO is understood as the set of strategies of the company to manage socio-political risks and opportunities that could impact the company's ability to operate; and the set of agreements with the external stakeholders to ensure a contribution to their sustainable development. These risks and opportunities relate to impact management, quality of engagement and procedural fairness. In summary, the SPLTO is the framework of the relationship between the mining operation and the external stakeholders for mutual benefit during the operation and post closure.

For instance, in the updated version of their social policy, the Social Way, Anglo American sets guiding principles for stakeholder engagement – such as dialogue-based, strategic, proactive, clear and direct, inclusive, ongoing, shared and understood internally – establishing clear rules for building a mutually beneficial relationship with those affected and those that can affect an operation. The contribution to SDG 11 of such a practice is, first and foremost, the recognition that there are affected parties and not only parties that can affect the operation. Such stakeholder engagement practice can also encourage communities to organise themselves, or to strengthen their existing organisations in order to interact with the company in more equal conditions, so they are able to set the terms of the relationship; this enables them to create win–win agreements and to obtain benefits from the operations, besides allowing them to be informed in a transparent, well-communicated and timely manner about the operations, impacts and benefits of a specific operation.

Additionally, when done in line with international standards and best practices, stakeholder engagement plans should identify vulnerable stakeholders, indigenous peoples and other marginalised groups, in order to design specific mechanisms to ensure their voices are heard, and their opinions and needs are included in social programs.

Due to the increasing scrutiny by civil society of mining companies, they are now more open and are creating different accountability mechanisms, which gradually create spaces for locals to participate in decision-making.

These practices could be a contribution to SDG 11 as they can enable communities and their social organisations to find common interests, to empower themselves, to raise their voices when interacting and negotiating with mining companies, and also to ensure the companies implement culturally and environmentally adequate impact and risk prevention and mitigation measures. While difficult to measure, this empowering dynamic can facilitate movement of communities from very basic levels of self-preservation towards a more inclusive approach and way of living.

Nevertheless, even stakeholder engagement may pose risks to SDG 11, as will now be presented.

## Risks and Impacts of Mining for SDG 11

Even though, as already presented, mining projects and operations have natural synergies with fulfilling SDG 11, and also make contributions to achieving some of the SDG targets, these potential contributions also pose risks to and have adverse impacts on the community´s capacity to be or become inclusive, safe, resilient and sustainable.

Mining-induced displacement and resettlement can be considered one of the highest socio-cultural and economic impacts of mining, as it affects access to land, livelihoods, social networks and cultural practices, and the adaptation of communities to their new environment. Restoration of livelihoods and cultural networks is a long-term process that might not be fulfilled, due to conditions such as: lack of capacity of the receiving communities; relocation into less productive land; weak local government that fails to insert new communities into their public and social services infrastructure and systems; increased dependency on the mining operation, the LRP and the mine during the process, among other conditions.

Involuntary resettlement can be triggered by unplanned or unidentified environmental impacts. Such is the case of the El Hatillo, Boqueron and Plan

Bonito communities in northern Colombia, where a ruling by the Ministry of Environment, Housing and Development obliged Drummond, Vale and Prodeco to resettle these three communities due to air pollution caused by the coal dust produced at the three coal mining operations in the region operated by each of these companies. The ruling was made in 2010, and, to date, the resettlement has not been completed. The uncertainty around the RAP and LRP, and around the negotiation process and the responsibilities of the involved parties (government, companies, communities), has had a psychological effect on the communities and individuals; it has created conflict and put people's lives on hold, which has impacted their resilience and the sustainability of their livelihoods, not to mention the impacts on the environment and on health, especially of the most vulnerable such as children and the elderly.

Besides resettlement, mining can impact livelihoods as it can cause inflation. In tribal societies that do not have a cash-based economy, the payment of wages might seem to be an economic opportunity but in reality it may have not only a cultural impact but it can decrease the capacity of employees to have a sustainable livelihood, as it forces them to buy products instead of producing them independently of the mine.

Stakeholder engagement strategies and activities, that are intended to create trust and a fair relationship between companies and communities, also pose risks to, and impact on, inclusiveness and a community's ability to make decisions independently, as mining company's engagement strategies are usually implemented as one-way information processes: they tend to involve empowered and powerful stakeholders; are limited when including vulnerable and marginalised people; and community opinion is not usually included in a company's decision-making, leaving communities in a state of frustration, and discouraging participation and inclusion.

Furthermore, as the interactions are usually with politically weak communities, which lack social organisation and planning capacities, stakeholder engagement can lead to conflicts within and between communities, competition for social investment benefits, and conflict between communities and civil society organisations that are pro or against mining.

Innovation and less environmentally impacting operations can have adverse effects on sustainable development and resilience. For instance, automation, replacement of fleets by electric vehicles, the operation of waterless mines and the implementation of other technologies, can impact community access to land and resources, besides reducing the needs of the workforce and the number of jobs available to locals. New technologies aim to increase the productivity of mines, which means longer life for assets, and the capacity to extract more and different types of minerals in the existing mines, which can mean the generation of more waste, higher dewatering impacts, higher dust pollution, among other impacts that are to be identified. Additionally, automation demands different skills that might not be available locally, reducing the opportunity of access to jobs.

The impact of mining on community safety can also be high.

> In 2018, the 26 members of ICMM recorded a total of 50 fatalities [...]. [However], those numbers are clouded by inconsistencies in how miners report their numbers. For instance, the ICMM recommends that its members count deaths of employees killed

> transporting resources, like truck drivers hauling mine production. However, members including Anglo American PLC, Barrick Gold Corp., and BHP say they don't always, for different reasons.*

In some cases, community fatalities are not reported, as they are outside the mining lease or the direct area or influence of the operation. So, while difficult to measure, as statistics of the safety risks are not completely reliable, mining poses a risk to community safety due to increased traffic, waste management, water and air pollution, competition for resources and emergencies such as tailings, dam collapses and other issues related to the operations.

Sustainable peace-building has become relevant in various countries and is especially relevant for SDG 11 and a community's resilience. The Colombia case was mentioned earlier, and Glencore reports its contribution to peace-building through social investment programs. However, the company has had allegations of violation of Human Rights made against it and linkages with illegal groups, also in Colombia, implied† which, if they turn out to be true, could undermine not only the company's contribution to peace-building, but the community's ability to live free of violent conflict.

Furthermore, mining companies and, in general, the private sector, does not yet have a proper understanding of their role in peace-building, due to the business model that will be now be discussed.

## LESSONS LEARNED

### The Business Model

Most major mining companies are committed to investing effort and resources in specific SDGs, and report against SDGs in their annual Sustainability Reports, SDG 8, 3 and 6 being most frequently included. Some major mining companies report that they contribute to SDG 11 by partnering with communities and governments for community sustainable development and resilience.

Nevertheless, some of these efforts are linked to the SDGs, and SDG 11 in particular, only for reporting purposes, that is to say, the links are not made in the planning process, and the CSR, SED and stakeholder engagement activities are not aimed at making a systemic contribution to community sustainable development, safety, resilience and inclusion; these programs are mostly individual, separate actions, that, combined, can contribute to SDG 11.

Additionally, the key drivers for mining companies to invest in sustainable communities are mostly related to creating an enabling and stable environment for the business. These include maintaining social support to the operation (misunderstood as having and maintaining the SPLTO), ensuring legal compliance, managing operational and financial risks, creating a positive reputation – all of which ultimately contribute to ensuring the sustainability and expansion of the business.

---

* www.wsj.com/articles/the-hidden-deaths-of-mining-11577825555?mod=searchresults&page=1&pos=3

† www.business-humanrights.org/es/colombia-informe-relaciona-a-prodeco-y-drummond-con-%E2%80%9Cparas%E2%80%9D

The majority of these social programs and contributions present three key issues: (a) they are not consistent with the corporate policies and standards and are mostly isolated and dispersed efforts; (b) the majority of them are focused on managing risks to the business; and (c) they still obey an "enclave" business model.

a) Inconsistency with corporate policies and standards: As is shown in the mining companies' sustainability reports, stakeholder engagement, Corporate Social Responsibility and socio-economic development programs can be seen as isolated and disperse efforts that i) lack structure, a strategic and long-term view, and a purpose that could help catalyse SDG-aligned regional development; ii) are planned and implemented in isolation from each other; iii) are planned and implemented separately from the life of asset and operational plans; iv) are not aligned with corporate-level policies, standards and mandates, but obey the site's immediate needs; in fact, v) these programs are even, in many cases, isolated from community priorities, as they are usually not planned in consultation with the beneficiaries.

   So, what has been called natural synergies and mining contributions to SDG 11 could also be called residual effects of "doing business as usual". This might come from a perspective that only focuses on ensuring an enabling environment for the business, and not on being part of the local and regional socio-economic dynamics.
b) Managing risks to the business: Mining industry social programs and activities are shown to be mainly aimed at preventing or responding to: risks to and impacts on the business; ensuring operational continuity; and to preventing any social incidents that can affect production, reputation or have legal implications (blockades, demonstrations, strikes, etc.). This approach often results in "quick wins" with only a short-term impact, and not a significant contribution to community sustainability, safety and resilience.
c) The "enclave" model: Most mining companies still operate in a way that does not fully integrate into the local/regional socio-economic system and dynamics. This approach leads to a social intervention focused mainly on the area of influence of the operations or projects, and to investment in short-term programs. It also leads to an individualistic approach that fails to partner with or invest in the diverse stakeholders at a regional level, including other mining companies. If mining companies were to become active actors of the local and regional dynamics, had an active role in the regional planning and brought their know-how to their partnerships with governments, other mining companies, other industries and civil society organisations, they could be catalysts for regional development and, as a consequence, would have a profitable business. Right now, the business model is the other way round: the purpose of mining companies is to have a profitable business and they expect this will have a positive effect on local economies. Herein lies the limitation of the current business model.

This business model not only contributes to SDG 11, but can, in fact, exacerbate vulnerability and induce other vulnerabilities in communities, which constitutes a hindrance to moving towards a local and regional sustainable development.

## Challenging Regional and Local Socio-Economic Context

There are factors that are external to the mining operations and which challenge their contribution to SDG 11. Mining usually operates in remote, isolated locations where local communities experience challenges to accessing basic public services (potable water, sanitation), social services (health, education), have high levels of poverty, are usually politically weak and are not integrated into the regional development dynamics.

Such communities have high levels of vulnerability, limiting their ability to recover from the impacts of mining operations. In this context, community priorities are at the self-protective level, to ensure a basic standard of living and a livelihood that secures access to food.

As a consequence, communities and even local and regional authorities, lack the capacity to plan for their sustainable development in the long term, and lack the empowerment to partner with mining companies to implement actions and plans that contribute to their sustainable development, inclusion and resilience.

## The Role of International Initiatives

There is an increasing number of international initiatives regarding mining's contribution to SDGs.* Even though these initiatives may constitute a key driver for mining companies about their role in regional and local sustainable development, inclusion and resilience, most of these initiatives do not constitute a requirement for financing projects (such as the IFC Performance Standards); as a consequence, they are not always fully implemented.

Despite the positive reputational impact that aligning with the SDGs and committing to their achievement creates for mining companies, the SDGs are not yet embedded into companies' policies, standards, processes and actions on the ground.

International organisations and mining councils have a critical role to play and can start by including measurable contributions to SDG 11 into their guidelines and standards for their members, and by assessing operations at the local level with clear criteria.

Furthermore, requiring alignment and strict indicators on the achievement of SDGs, especially SDG 11, which seeks community safety, inclusion and resilience, as a condition for financing projects and operations, could help embed good practices further, and to put them in place through holistic strategies of integration in the local

* Some international initiatives that could contribute to aligning businesses and the achievement of SDG 11 are: UNO Business and Human Rights Guiding Principles; Voluntary Principles on Security and Human Rights; ICMM statements and best practices; Initiative for Responsible Mining Assurance (IRMA); Kellog Innovation Network – Mining Catalyst (KIN Mining Catalyst); UNDP APELL Process (Awareness and Preparedness for Emergencies at Local Level); IFC Performance Standards, among others.

dynamics, through partnerships with local and regional governments, other mining companies, other economic sectors and civil society, in order to co-create long term sustainable development plans that ensure independence from mining.

### Way Forward

The mining sector's purpose must continue to evolve beyond creation of value for shareholders and impact management, towards a more active role in society, as catalysts of local and regional sustainable development, community empowerment and social evolution. This will require companies and the sector at large to structure mechanisms for incorporating community perspectives into decision-making; to integrate into regional dynamics through partnerships with the communities and different stakeholders; and to help create the basis for communities to evolve and move away from self-subsistence towards a culturally relevant vision of sustainability.

Mining companies, with their know-how, global scope and capacity to influence governments and international organisms, should evolve towards creating regional development that, as a consequence, creates an enabling environment for their operations.

## CONCLUSION

The contribution of mining to SDG 11 is still limited to diverse activities and intervention in the area of influence of the operations. The key drivers are mostly managing operational risks and impacts in order to create a stable enabling environment for the business. The current business model is predominantly based on short-term goals.

When mining companies implement good practices and high standards across the business, and in all the sites and operations, in a holistic and systemic way, their contribution to SDG 11 will be more significant, and will trigger and initiate social cohesion, impelling communities to grow, empower themselves, raise their voices, participate in decision-making, and own their path towards sustainable development.

## CASE ANALYSIS 1

## COLLABORATIVE REGIONAL DEVELOPMENT: KEY ENGINE FOR ANGLO AMERICAN´S FULFILMENT OF ITS BUSINESS PURPOSE

**Laura Garcia Jaramillo**

Anglo American is a mining company with operations across the globe. Its portfolio includes diamonds, copper, PGMs, coal, iron ore, nickel and manganese. The company has a clear purpose, which goes beyond the creation of value for its shareholders, and is taking a broader and more active role in regional development. Anglo American´s purpose is to "reimagine mining to improve people´s lives", and the way to live up to it is through Future Smart

Mining, an "innovation-led approach to sustainable mining". One component of Future Smart Mining is the Sustainable Mining Plan (SMP), which is based on three pillars: trusted corporate leader, thriving communities and healthy environment. All sites across the globe will develop tailored five-year plans in order to meet the global stretch of each pillar.

### Pillars and Components of Anglo American's SMP and Its Contribution to SDGs

| Pillar | Key Components | Stretch Goals (to be achieved by 2030) | Contribution to SDGs |
|---|---|---|---|
| **Trusted Corporate Leader** | Accountability | Establish multiple levels of open and accountable dialogue leading to greater mutual trust<br>Widespread recognition of the benefits that responsible mining can bring, a collaborative openness about the challenges and a more consensual working relationship between Anglo American and society | 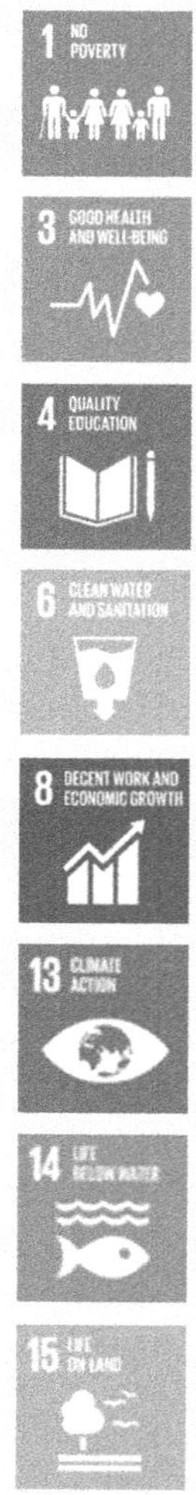 |
| | Ethical value chains | Responsible sourcing standard fully implemented across Anglo American<br>Advocate for the development of a common responsible sourcing standard for the mining industry | |
| | Policy advocacy | External recognition of Anglo American´s leadership on policy advocacy<br>Strong levels of engagement in policy debates | |
| **Thriving Communities** | Education | Schools in host communities to perform within the top 30% of state schools nationally | |
| | Health and well-being | SDG 3 targets for health to be achieved in our host communities | |
| | Livelihoods | Five jobs created/supported offsite for every job on site | |
| **Healthy environment** | Biodiversity | Deliver NPI on biodiversity across Anglo American | |
| | Climate Change | Reduce the net GHG emissions by 30%<br>Improve energy efficiency by 30% | |
| | Water | Reduce freshwater withdrawal of regions by 50% | |

A second key component of Anglo American's SMP is Collaborative Regional Development, an approach to sustainable development that seeks to integrate the operation into the regional socio-economic dynamics, and incubate and catalyse development of the region through spatial analysis, partnerships and a strategic, long-term approach.

After recognising the cumulative socio-economic and environmental impacts of mining, the high economic dependence communities have created on mining, the adverse effects of mine closure, and living up to Anglo American´s purpose, CRD looks at a 30-plus-year sustainable and inclusive economic development that creates opportunities away from mining and leverages regional potential in all its extents.

### Key Elements of CRD and Its Contribution to SDGs

| Elements | Implementation | Contribution to SDGs |
| --- | --- | --- |
| **Spatial Analysis and Planning** | Identification of socio-economic development opportunities with the greatest potential in a region. Spatial analysis respects and builds on critical natural resources and features, identifies socio-economic vulnerabilities and opportunities to introduces more resilient systems through spatially targeted interventions. This analysis includes the opportunities and maximum environmental resources levels (access to water, social services, public services, etc.). |   |
| **Partnerships** | With a broad range of stakeholders (private sector, government and public sector, communities, entrepreneurs, academics, NGOs, faith groups, etc.), that together can catalyse regional development opportunities into action. CRD establishes mechanisms to ensure the parties involved are listened to, co-create long-term solutions, and have collective impact. This also means that communities will not only be beneficiaries of the programs to be implemented but will also participate in the wealth they create. |   |
| **Strategic Long-Term Approach** | Building the foundations for long-term, sustainable development in the host regions far beyond the life of the mine. |   |

### Implementation of the SMP and CRD: Challenges and Lessons Learned

The SMP was launched in 2018, and sites are in the process of setting their 2020 goals and preparing their five-year implementation plan. CRD started being implemented in 2017 in South Africa, at the Limpopo Province, where the largest platinum operations are located, and has continued in 2019 in Peru, where the Quellaveco copper mine is currently being built.

Lessons learned according to Anglo American employees in charge of its implementation can be summarised as follows:

- **Integration:** Cross-functional integration and collaboration are challenging, as the culture of most mining companies is to operate in silos. The perspective shift brought by the SMP and CRD is being integrated into the policies, procedures and the operational processes at site level. This entails transforming the culture of the whole organisation, and the way socio-economic development has been done in the past, which is a long-term process that has just started.
- **Capacity, capability and resourcing at site level:** The site social performance and socio-economic development teams are mostly focused on ensuring the socio-political context is enabling for the business, on managing any potential social crisis and on providing short-term solutions for the stakeholders of the area of influence. The availability of the teams to implement CRD is reduced, and the capacity for implementing this innovative approach is to be built. Furthermore, operations are mostly focused on "quick wins" that can be leveraged to sustain the social licence to operate, and 30-plus-year programs become less attractive both to operation managers and to external stakeholders, which have more immediate needs to resolve.
- **Legacy issues:** Some of AA's operations, such as the platinum ones, have been there for decades. Others, such as Quellaveco, are being built in areas where other mining companies have been operating under different standards. The legacy issues are still being approached and resolved. At some sites, the minimum social performance requirements have not been met yet, and basic community needs are not solved. The tension between the past impacts that are being currently experienced and the potential future benefits creates difficulties for prioritisation of programs for sustainable development.
- **Partnerships, co-creation and co-implementation:** Anglo American is setting up the platform for the multi-stakeholder design and implementation of CRD. The capacity (financial, availability of resources, capability) and interests of all the stakeholders are not the same. Finding common ground, levelling the ground, and establishing the terms of reference for partnerships is challenging and is a trial-and-error process. Additionally, alignment of timing and governmental cycles of planning and implementing, not to mention elections, adds high uncertainty and instability to the implementation of the programs.
- **Deep transformations:** The results of spatial analysis might show that the potential of the region is different and even opposite to what is embedded in the region´s culture and identity, and far from current political interests. This entails a process of questioning the entire regional socio-economic dynamics, which becomes a source of conflict among the different partners involved.

- **A needed cultural change:** SMP and CRD are innovative approaches that will contribute directly and indirectly to achieving the SDGs in some of the most challenging geographies. Implementation has just started, and companies, communities, governments, academia, NGOs and other sectors are in the process of opening themselves to having meaningful conversations and supporting each other in this initiative. The sense of urgency is present, although there is still a long way to go.

## CASE ANALYSIS 2

## KAHAMA MINING CORPORATION'S HOUSING AND SOCIAL DEVELOPMENT PLAN

*Tanzania*

**Dr. Janet Fishlock**

*Dr. Janet Fishlock was a senior consultant with planning Alliance at the time of this writing and spent 18 months in Tanzania working with KMCL to develop the housing and social development plan and approach. Currently she is a freelance consultant and steward with the Social Practice Forum (SPF).*

This case study examines the partnership approach underpinning the Housing and Social Development Programme of Kahama Mining Corporation Limited (KMCL). It was designed and implemented at KMCL's Bulyanhulu gold mine in Tanzania in early 2000. It illustrates the contribution mining can make to building sustainable communities.

The Tanzanian government at that time was intent on encouraging foreign investment, particularly in the mining sector, and Barrick Gold, a major Canadian mining company, was planning to establish a series of mining operations in the region. This included the underground mine in Bulyanhulu. Expectations that the mining industry would make a positive contribution to local community development were high, particularly in the Bulyanhulu area where most households relied predominantly on subsistence agriculture; health infrastructure was poor and education levels amongst the lowest in the country.

KMCL required a significant workforce for their underground operation. Although they would need to rely on some expatriate personnel, they were committed to hiring from within Tanzania and reducing their expatriate workforce by 70% over five years. Attracting and retaining quality personnel from across the country was going to be a key challenge. KMCL commissioned planning Alliance (pA), a Canadian-based consulting company with international experience, to assist them in developing and implementing a housing and social development plan that would both improve local infrastructure in the area, and provide quality housing for their workforce. The development of the plan involved extensive consultation and negotiation with KMCL management, mine workers, area residents and local government. It was key that

KMCL's social development plan align with the priorities outlined in the District Development Plan being established at that time.

The housing scheme that evolved provided mine workers with the opportunity to live with their families in well-constructed homes they could own over time, in local communities near the mine. This would enable residents of these communities to benefit from the improved infrastructure and services (i.e. roads, electricity and water) associated with the housing scheme. Community education and health programs that KMCL developed in collaboration with local partners (i.e. CARE Tanzania and AMREF) were also part of the housing and social development plan. The goal was to improve the quality of education and health services available to mine workers, their families and existing residents.

While KMCL's original intention was that implementation of the social development plan would be shared equally amongst the various partners, community and government capacity and resources were limited. Consequently, the company was compelled to take a leadership role in providing the bulk of the funding as well as contributing contract and project management skills. Despite this, the partnership principles of joint design and decision-making, shared risks and responsibilities and a pooling of resources and competencies were adhered to through the establishment of a series of multi-party committees for each of the main program components. Steering committees included representation from KMCL management, mine workers, residents, NGOs and local government. And, in addition to overseeing implementation of the various programs, these committees planned for the transferring of responsibility for infrastructure management and maintenance to communities and local government.

Key benefits resulting from the housing and social development plan included strong community–company relationships and an improved quality of life for KMCL employees and many area residents. Close to 600 houses were constructed as part of the housing scheme, with access to a reliable water supply for both mineworkers and the 5000 residents of Bugarama and Ilogi villages. The five-year education improvement program led by CARE Tanzania in partnership with local schools and the district education department, resulted in increased rates of enrolment, attendance and ultimately graduation at the primary school level in the area. Health security was improved through increased community knowledge and understanding of HIV/AIDS and malaria and the adoption of improved health practices to prevent the spread of these diseases. KMCL's social development plan also strengthened local governance, by assisting the District Council in implementing most of its development plans for the area around the mine, despite limited government funds.

KMCL's housing and social development plan was featured as a case study by the Secretariat of the Natural Resources Cluster (NRC) of Business Partners for Development (BPD). This major research program, which ran from 1998 to 2002, aimed to enhance the role of oil, gas and mining corporations in community and regional development. In 2004, Michael Warner and Rory Sullivan published it in a book entitled *Putting Partnerships to Work: Strategic Alliances for Development between Government, the Private Sector and Civil Society* (first published by American Planning Association and then Routledge).

## BIBLIOGRAPHY

Business and Sustainable Development Commission. http://businesscommission.org/.

Business Fights Poverty. https://businessfightspoverty.org/articles/embedding-the-sustainable-development-goals-into-business/.

Corporate Human Rights Benchmark. 2019 Report. https://www.corporatebenchmark.org/sites/default/files/2019-11/CHRB2019KeyFindingsReport.pdf.

Department for International Development (DFID). 1998. Sustainable Livelihoods Guidance Sheets.

DFID. 1999. *Sustainable Livelihoods Guidance Sheets*, Department for International Development: London.

Good Practice in Emergency Preparedness and Response. 2005. ICMM. https://www.icmm.com/website/publications/pdfs/health-and-safety/good-practice-emergency-preparedness-and-response.

ICMM Integrated Mine Closure Plan. https://guidance.miningwithprinciples.com/integrated-mine-closure-good-practice-guide/.

ICMM Making a Positive Contribution to Sustainable Development. https://metals.miningwithprinciples.com/.

ICMM Social Progress in Mining Dependent Countries. https://www.icmm.com/social-progress.

IFC Projects and People: A Hand-Book to Address Project-Induced in-Migration. https://www.ifc.org/wps/wcm/connect/topics_ext_content/ifc_external_corporate_site/sustainability-at-ifc/publications/publications_handbook_inmigration.

Johnston, S. & D. Kemp, 2019. *Communities and Social Performance in the Global Mining Industry: A Snapshot Study*, Centre for Social Responsibility in Mining, Sustainable Minerals Institute, The University of Queensland: Brisbane.

Kellogg Innovation Framework. 2012. Reinventing Mining: Creating Sustainable Value. Introducing the Development Partner Framework. http://www.kinglobal.org/uploads/5/2/1/6/52161657/pb_kin_dpf_final_12_4_5mb.pdf.

Kellogg Innovation Framework. http://www.kinglobal.org/mining-catalyst.html.

Land Acquisition and Resettlement: Lessons Learned. 2014. ICMM. https://www.icmm.com/website/publications/pdfs/social-and-economic-development/9714.pdf.

MacDonald, Alistair, Patricia Kowsmann, & Krishna Pokharel. 2019. The Hidden Deaths of Mining. *The Wall Street Journal*. https://www.wsj.com/articles/the-hidden-deaths-of-mining-11577825555?mod=searchresults&page=1&pos=3.

Mining Technology. 2019. The Top Ten Mental and Mining Companies by Revenue 2018. https://www.mining-technology.com/features/the-top-ten-metals-and-mining-companies/.

Owen, J., G. Muriuki, & D. Kemp, 2018. *Livelihoods, Food Security and Mining-Induced Displacement and Resettlement*, Centre for Social Responsibility in Mining (CSRM), The University of Queensland: Brisbane.

Owen, J.R. & D. Kemp 2015. Mining-Induced Displacement and Resettlement: A Critical Appraisal. *Journal of Cleaner Production*, 87(1), pp. 478–488.

Ruling of the Ministry of Environment, Housing and Territorial Development of Colombia on the Resettlement of Plan Bonito. 2010. (Sentencia Ministerio de Ambiente, Vivienda y Desarrollo territorial reasentamiento Plan Bonito). http://www.minambiente.gov.co/images/normativa/Otros/Autos/2012/auto_0457_2012.pdf.

SDG Compass: The Guide for Business Action on the SDGs. https://sdgcompass.org/wp-content/uploads/2015/12/019104_SDG_Compass_Guide_2015.pdf.

http://www.livelihoodscentre.org/documents/20720/100145/Sustainable+livelihoods+guidance+sheets/8f35b59f-8207-43fc-8b99-df75d3000e86.

GOAL

# 12 Ensure Sustainable Consumption and Production Patterns

*Damien Giurco*

## CONTENTS



## A SUMMARY OF THE RELATIONSHIP OF MINING TO SDG 12: SUSTAINABLE PRODUCTION AND CONSUMPTION

| | |
|---|---|
| Key stakeholders | • Mining companies<br>• Coordinating/certifying organisations, e.g. ResponsibleSteel™, Aluminium Stewardship Initiative, CERA: Certification of Raw Materials<br>• UN Environment – International Resource Panel<br>• World Resources Forum<br>• One Planet network – a hub for SDG 12<br>• IRMA – Initiative for Responsible Mining Assurance<br>• Investors<br>• Manufacturers/end-users of metals<br>• Governments, via regulation and procurement<br>• Civil society organizations |
| Direct mining contribution/Drivers | Drivers for the mining industry come from<br>• Investors/shareholders – seeking improved performance and reduced social and environmental risk<br>• Community – seeking responsible production and the miner as a good corporate citizen<br>• End-users of metals – whilst still growing in profile, demand, for example from auto-makers for materials with a known and low environmental and social footprint is increasing; initiatives such as ResponsibleSteel™ are seeking to generate a market premium from responsible production |

| | |
|---|---|
| Contribution to SDG | Mining contributes to the SDG by providing metals for use; however, it can contribute to the deterioration of the goal as well through impacts of production. Places where metals are consumed often differ from where they are mined. As metals are demanded, including for low-carbon technologies required for responsible consumption and sustainable development, the production impacts in supplying countries must be recognized and better monitored and improved, including to better support the community development in the country through mining. |
| Key regulations and public policies | • UK Modern Slavery Act 2015<br>• Australian Modern Slavery Act 2018<br>• China's National Plan on Implementation of the 2030 Agenda for Sustainable Development<br>• Malaysian National Blueprint on Sustainable Consumption and Production |
| Corporate policies and private programs | Drive sustainability – aiming to drive sustainability in the global automotive supply chain by integrating sustainability into procurement https://drivesustainability.org/<br>Green star – voluntary sustainability rating systems for buildings in Australia run by the Green Building Council of Australia, supporting the use of ResponsibleSteel™ https://new.gbca.org.au/green-star/<br>Open government partnership – working to transform how government serves its citizens www.opengovpartnership.org/<br>Extractive Industries Transparency Initiative – standard to promote the open and accountable management of oil, gas and mineral resources https://eiti.org/<br>TCFD – Task-force on Climate-related Financial Disclosures www.fsb-tcfd.org/<br>ISO 204000 – Sustainable procurement – Guidance www.iso.org/standard/63026.html<br>Coalition for energy efficient comminution www.ceecthefuture.org/<br>IRMA – Initiative for Responsible Mining Assurance https://responsiblemining.net/<br>CERA – Certification of Raw Materials www.cera-standard.org/ |

In relation to the goal of sustainable consumption and production, the mining industry has historically been focused on sustainable production, beginning with reducing environmental and social impacts of production, and less on sustainable consumption, namely considering the ultimate rates of metal consumption and the sustainability of patterns of consumption. This has begun to change somewhat in recent years with the increasing focus on the circular economy which encourages a systems view of production and consumption. Mining companies now need to reimagine their role and value proposition in a low-carbon future as a supplier of resources and services in a circular economy, alongside an increasing role played by recycling, reuse and remanufacturing. Indeed, some recyclers consider themselves miners of urban stocks of metals from discarded products and infrastructure.

## ANALYSIS OF THE CONTRIBUTION OF MINING TO RESPONSIBLE CONSUMPTION AND PRODUCTION

The 10 Year Framework on Responsible Consumption and Production sits at the heart of SGD Goal 12 and is now championed by the One Planet Network (www.oneplanetnetwork.org/). It has a focus on (i) sustainable buildings and construction, (ii) sustainable public procurement, (iii) sustainable tourism, (iv) sustainable food systems, (v) sustainable lifestyles and education and (vi) consumer education. Whilst mining is not explicitly mentioned in the One Planet Network's current five-year strategy, as a supplier of materials required for a diverse range of sectors and technologies, responsible mining is important for achieving the goal and a greater focus on the sector's possible contribution to the SDG would be beneficial. Also relevant to SDG 12 is the reform of taxation and harmful subsidies to polluting sectors – this is of particular relevance to fossil fuel subsidies going to coal mining.

By contrast, if one begins with a mining sector view, the publication "Mapping Mining to the Sustainable Development Goals: An Atlas" (2016) highlights the issues of materials stewardship, circular economy and minimizing waste as most relevant for SDG 12, with a focus on minimizing resource use and waste and incorporating life-cycle thinking into core business. It rightly notes the important contribution that mining – in enabling the provision of metals – makes to supporting sustainability and responsible consumption across a range of sectors from transport, electricity, construction and communications.

Things have improved over recent decades since the Mining, Minerals and Sustainable Development Initiative which concluded in 2002 that technological improvements to make mining and the provision of metals more efficient should be pursued: however, increasing the tonnes of production from a mine often remains an overriding goal, even when changed practices at the mine could reduce subsequent refining impacts, for example by optimizing grade/recovery curves.

Chains of custody schemes have grown, driven by end-user demands from government and industry for responsibly sourced materials, although it has recently been acknowledged that there are limits to traceability and the results we can expect from this measure alone (Soovacol et al. 2020).

At the level of the mine site, there are noted cases of improving water and energy efficiency and frameworks for governing water have improved.

To get a better take on the role of mining's contribution to Goal 12, the conversation needs to be engaged in a systems view, analyzing not only new mining technology but the contribution of minerals to new technologies used in society through a systems lens. To assess sustainable consumption, the role that individual, collective and institutional behaviours play in building and improving systems must also be considered.

## LESSONS LEARNED

In the pursuit of responsible consumption and production, a number of lessons can be noted and are outlined as follows.

*Manage water for both drought and floods* – with the advancement of climate change, planning for responsible water management requires not only considering more efficient water use during times of drought and water scarcity, but managing more severe floods. The lesson in some locations is that an adaptive planning approach is required that considers both realities, that is, during operation and also in planning for closure and safe tailings storage.

*Resource-energy nexus is important* – when considering sustainable development, including a transition to cleaner energy sources, the impacts of resourcing the minerals and metals required for this infrastructure and sustainable development is now recognized as an important issue. Renewable energy continues to be adopted in mining and minerals processing both for economic reasons (as the capital costs of renewables have fallen and interest rates are low) and to be part of supporting low-carbon supplies demanded by end-users. Looking ahead, as renewable electricity is less easy to transport over long distances than for example coal or oil, a competitive advantage may emerge in processing metals closer to where they are mined, if ample low-carbon energy is available, rather than closer to markets.

*Responsible supply chains* have begun to be more formally codified, such as via the Aluminium Stewardship Initiative, The Copper Mark, ResponsibleSteel™, Responsible Cobalt Initiative. As multi-stakeholder processes, they take time to effect change, but more is being learned, including across commodities about how to make the initiatives ambitious enough to make a difference whilst trying to bring most along so the impact of the change extends beyond front-runners. The role of sustainable procurement, by both companies and governments, is increasing as a driver for responsible supply chains.

*Life of mine impacts* – responsible supply requires sustaining development with the community where mining takes place. The jobs being provided by mines vary by mine location, but are changing with increasing automation. In addition to social and community impacts, the environmental footprint of the mine through to planning for closure is under the spotlight, after many high-profile tailings dams failures. The lessons learned here include that it may be worth adopting new processing approaches (e.g. dry tailings or even in-situ mining). Ocean mining and asteroid mining have also received increased attention; however, the environmental costs and governance for ocean mining remain contested, and questions over techno-economics and governance for asteroid mining are yet to see it occur.

*Closing the loop in a circular economy* – as the scale and impacts of future material use-patterns are recognized, the circular economy is being increasingly encouraged by business and governments to focus on extending resource use through reuse, remanufacture and ultimately recycling. For mining, this leads to opportunities for improved onsite management of consumables, vehicles and equipment and also a reimagining of the role of mining – relative to secondary sources – for future material supply in a circular economy. Whilst recycling often has a reduced environmental impact, there are still social and environmental issues (e.g. labour conditions for ship breakers recovering iron).

The circular economy needs to couple with responsible consumption rates and requires improved approaches to design for circularity. What may look efficient for a once-through linear economy (e.g. light weighting) may be counterproductive for a multi-use circular economy where durability for reuse is a more important attribute in a product casing or packaging or where the higher contained material value underpins the economics of recycling. A further lesson learned is that the processing and thermodynamics of recycled materials also need more scientific study as the metallurgy of processing metals which have been assembled in products is different from the process of separating metals in naturally occurring ores.

*Reimagining business* – simply promoting circular flows of resources – even if coupled with low-carbon energy sources – may well be insufficient for configuring systems of responsible consumption and production. Total demand, driving the rate of materials circulating and the impacts arising, must also be altered to fit within planetary boundaries.

Here, a shift in mindset for the minerals industry, from extracting and selling commodities, to providers of metal and material services, is one opening for more profound transformation. This would be likely to require a change in business models, from resource sales to resource stewardship, for example, by commodity leasing over selling (and with more of a long-term stewardship ethos in the leasing than the precious metal leasing that is currently available to hedge market price risk).

At the time of writing, coronavirus is spreading across the globe and is overturning business-as-usual practices. Whilst holding businesses and societies together during this period will be a formidable challenge, it brings into focus a discussion of our values as a global society. The experience demonstrates that business-as-usual can be radically and quickly altered – as will need to occur in a future that is adapting to being carbon negative.

In the reimagining of considered, circular consumption, and the rebuilding of resilient, responsible production, the mining industry has a unique opportunity to rearticulate its role in bringing sustainability to our communities, countries and our planet.

## CASE ANALYSIS

### URBAN MINING AND CIRCULAR ECONOMY: PAVING THE WAY TO SDG 12 AND A REPRIEVE FOR OCEANS FROM DEEP SEA MINING

**Helen Rosenbaum and Natalie Lowrey**

Next year, a little-known United Nations agency, the International Seabed Authority (ISA), is expected to open up the high seas to mining. The body, which is based in Kingston, Jamaica, is due to finalize its mining code: a set of regulations for exploiting the sea floor in international waters. The ISA has already completed regulations and recommendations for exploration. These have enabled it to grant 29 exploration licences in international waters (Aldred, 2019).

Whilst there is intense interest in the future financial returns that may be available through seabed mining, no commercial seabed mine has yet been established. Thus far, the industry remains a speculative and experimental activity driven in large part by commercial and geostrategic competition, as states consider ways to secure access to rare earth minerals that may become increasingly valuable.

Science about the risks of deep-sea mining is limited, despite the narratives spun by proponents such as mining company DeepGreen Resources. What is known, points to serious and irreversible impacts on marine ecosystems. (Simon-Lledó, Bett, Huvenne et al., 2019) These are already under stress and predicted by the IPCC to suffer further severe biophysical changes under business-as-usual scenarios with serious ramifications for human populations (IPCC, 2019).

The IPCC's predictions highlight the recklessness of pursuing DSM at this juncture – especially when alternative sources of minerals already signpost the route to SDG 12.

Research and development in urban mining, is well underway – the process of reclaiming compounds and elements from products, buildings and waste. A staggering 320 tons of gold and more than 7,500 tons of silver estimated to be worth $US21 billion is used annually to make personal computers, mobile phones, tablets and other electronic products worldwide (UN University, 2012).

There is an abundance of gold, silver, rare earths and copper in the waste generated by the disposal of these products. It is estimated that electronic waste contains precious metal "deposits" 40 to 50 times richer than the ores currently mined (Winter, 2013). Urban mining could be more lucrative than deep sea mining, would resolve intractable waste problems, and modelling has demonstrated its capacity to meet future global mineral demand (Teske et al., 2016).

The circular economy describes an economic system grounded in "cradle to cradle" product design, reconditioning, waste prevention and closed-loop production processes. In response to the global momentum towards this new paradigm, several companies have already put circular economy principles into practice. Apple declared in 2017 it would "stop mining the Earth altogether" and in 2019 announced it will include a component made of 100% recycled rare earths in its new iPhones (Jamasmie, 2019). The European Commission has introduced a circular economic framework (European Commission, 2014) and the European Parliament has called for a moratorium on deep-sea mining (European Parliament, 2018) until the need for it has been proven, as have prominent scientists, academics and civil society organisations (Our Common Heritage, 2018).

The way forward to SDG 12 is self-evident. We have a window of opportunity now to save our seas from the spectre of deep-sea mining and to create a future in which "dinosaur era" mining of virgin minerals will only be necessary to fill gaps in demand left by innovative production systems underpinned by urban mining and circular economy and their wins for society, the environment and investment.

## REFERENCES

Aldred, J. 2019. The future of deep seabed mining. *China Dialogue*. https://chinadialogueocean.net/6682-future-deep-seabed-mining/.

European Commission. 2014. Towards a circular economy: A zero waste programme for Europe. COM/2014/0398. https://eur-lex.europa.eu/legal-content/EN/TXT/?uri=CELEX:52014DC0398.

European Parliament. 2018. European Parliament resolution of 16 January 2018 on international ocean governance: An agenda for the future of our oceans in the context of the 2030 SDGs. 2017/2055 (INI). https://www.europarl.europa.eu/doceo/document/TA-8-2018-0004_EN.html?redirect.

IPCC. 2019. Intergovernmental panel on climate change. Special report on the ocean and cryosphere in a changing climate. https://www.ipcc.ch/site/assets/uploads/2019/09/SROCC-factsheet.pdf.

Jamasmie, C. 2019. Apple to include part made of 100% recycled rare earths in new iPhones. *Mining.com*. https://www.mining.com/apple-to-include-parts-made-of-100-recycled-rare-earths-in-new-iphones/.

Our Common Heritage. 2018. Joint letter noting international seas threatened by mining. Available with List of Signatories. http://ourcommonheritage.org/.

Simon-Llydó, E., B. J. Bett, V. A. I. Huvenne, K. Köser, T. Schoening, J. Greinert, and D. O. B. Jones. 2019. Biological effects 26 years after simulated deep-sea mining. *Scientific Reports* 9(1): 8040. https://www.nature.com/articles/s41598-019-44492-w.

Teske, S., N. Florin, E. Dominish et al. 2016. Renewable energy and deep sea mining: Supply, demand and scenarios. *Report Prepared by ISF for J.M.Kaplan Fund, Oceans 5 and Synchronicity Earth, July 2016.*

UN University. 2012. E-waste: Annual gold, silver "deposits" in new high-tech goods worth $21 billion+; less than 15% recovered. https://unu.edu/media-relations/releases/step-news-release-6-july-2012-e-waste-precious-metals-recovery.html.

Winter, C. 2013. Urban mining study to help develop an e-waste recycling industry. *ABC-Net*. https://www.abc.net.au/radio/programs/worldtoday/urban-mining-study-to-help-develop-an-e-waste/4751822?section=sa.

## BIBLIOGRAPHY

Ali, Saleem H., Damien Giurco, Nicholas Arndt et al. 2017. Mineral supply for sustainable development requires resource governance. *Nature* 543(7645): 367–372.

Ayuk, Elias T., Antonio M. Pedro, Paul Ekins et al. 2019. Mineral resource governance in the 21st century. Gearing extractive industries towards sustainable development. Summary for policymakers and business leaders.

Ciacci, Luca, Barbara K. Reck, N. T. Nassar, and T. E. Graedel. 2015. Lost by design. *Environmental Science and Technology* 49(16): 9443–9451.

Gasper, Des Amod Shah, Sunil Tankha, and S. Tankha. 2019. The framing of sustainable consumption and production in SDG 12. *Global Policy* 10(S1): 83–95.

Giurco, Damien, and Carlia Cooper. 2012. Mining and sustainability: Asking the right questions. *Minerals Engineering* 29: 3–12.

Mason, Leah M., Corinne Unger, A. J. Lederwasch, Hala Razian, L. E. Wynne, and Damien Giurco. 2013. *Adapting to Climate Risks and Extreme Weather: A Guide for Mining and Minerals Industry Professionals*. National Climate Change Adaptation Research Facility, Gold Coast.

Nansai, Keisuke, Yasushi Kondo, Damien Giurco, David Sussman, Kenichi Nakajima, Shigemi Kagawa, Wataru Takayanagi, Yosuke Shigetomi, and Susumu Tohno. 2019. Nexus between economy-wide metal inputs and the deterioration of sustainable development goals. *Resources, Conservation and Recycling* 149: 12–19.

Northey, S. A., G. M. Mudd, T. T. Werner, N. Haque, and M. Yellishetty M. 2019. Sustainable water management and improved corporate reporting in mining. *Water Resources and Industry* 21: 100104.

Oberle, Bruno, Stefan Bringezu, Steve Hatfield-Dodds et al. 2019. Global resources outlook 2019: Natural resources for the future we want.

Okwuosa, Innocent I., and Sharif S. Khalid. 2020. Contextualising accountability and responsible production and consumption in the extractive industries of West Africa. In: Ramutsindela M., Mickler D. (eds), *Africa and the Sustainable Development Goals*, pp. 177–185. Springer, Cham.

Reuter, Markus A., Antoinette van Schaik, Jens Gutzmer, Neill Bartie, and Alejandro Abadías-Llamas. 2019. Challenges of the circular economy: A material, metallurgical, and product design perspective. *Annual Review of Materials Research* 49(1): 253–274.

Sonesson, Casper, Gillian Davidson, and L. Sachs. 2016. Mapping mining to the sustainable development goals: An atlas. In: *World Economic Forum*, Geneva, Switzerland.

Sovacool, Benjamin K., Saleem H. Ali, Morgan Bazilian, Ben Radley, Benoit Nemery, Julia Okatz, and Dustin Mulvaney. 2020. Sustainable minerals and metals for a low-carbon future. *Science* 367(6473): 30–33.

Wiedmann, Thomas O., Heinz Schandl, Manfred Lenzen, Daniel Moran, Sangwon Suh, James West, and Keiichiro Kanemoto. 2015. The material footprint of nations. *Proceedings of the National Academy of Sciences of the United States of America* 112(20): 6271–6276.

# GOAL 13 Take Urgent Action to Combat Climate Change and Its Impacts

*Estelle Levin-Nally and Blanca Racionero Gómez*

## CONTENTS



## SUMMARY TABLE

| | |
|---|---|
| Key stakeholders | Local communities, mining companies, artisanal miners, processors and traders, regional and national governments, international bodies, mineral buyers and downstream end users |
| Direct mining contribution/drivers | Main drivers are growth in demand for:<br>• Low-carbon technologies<br>• Other technologies (e.g. ICT)<br>• Luxury goods<br>• Urban development<br>• Post-disaster reconstruction |
| Contribution to SDG 13 | Negative contributions<br>• GHG emissions (scopes 1, 2 and 3) from fossil fuel–based power generation for exploration, mining and mineral processing and transportation<br>• Land degradation, biodiversity loss, and deforestation related to mine and infrastructure development and secondary economic activities<br>• Displacement of communities<br>• Water, soil and air pollution<br>• Lobbying to stifle climate action<br>• Lack of coordination with other sectors |

| | |
|---|---|
| | Positive contributions<br>• Provision of minerals for low-carbon technologies and climate resilience (post-disaster reconstruction)<br>• Transferability to other sectors of climate-smart innovations to decarbonize mining and mineral processes<br>• Mines as sites for carbon sequestration<br>• Link between climate change and mining provides an entry point for a larger debate on socio-environmental standards |
| Key regulations and public policies | Examples include:<br>• United Nations Guiding Principles for Business and Human Rights<br>• UN Global Compact<br>• Proposed Green New Deals<br>• National regulations including post-COVID recovery strategies |
| Corporate policies and private programs | Examples include:<br>• The World Bank Climate Smart Mining Facility<br>• ACP-EU Development minerals program<br>• ICMM Sustainable Development Principles<br>• Climate Action 100+<br>• CDP<br>• RESOLVE's Salmon Gold Project<br>• Rio Tinto's Energy and Climate Change Commitment<br>• BHP Climate Investment Program |

## ANALYSIS OF THE CONTRIBUTION OF MINING TO SDG 13

The consumption of minerals has increased exponentially over the past century. Some studies predict that more metals will be produced in the next 30 years than over the past 100 (Vidal et al. 2017). Mining activities are expected to grow further, partly triggered by booming demand for technologies needed to transition to a low-carbon society, such as wind turbines, solar photovoltaics, electric vehicles, and batteries. For example, bauxite demand is expected to grow tenfold by 2030 compared to 2017, underpinned by shifts from plastic to aluminium packaging and from conventional to electric vehicles (Hogan 2019); lithium demand is forecasted to expand, mainly driven by ion-battery demand (Jamasmie 2018). The recycling of metals from the product's end of life can improve those metals' future availability by lowering the demand for virgin metals. Recycling rates, however, are unlikely to keep pace with the growth in demand (The World Bank 2017).

Other drivers of mining include the demand for luxury goods, especially in the emerging markets of India and China; new technologies not related to a low-carbon transition (such as mobile phones, computers, tablets, servers to support communication, automation, and the management and security of information etc.); new buildings and infrastructure due to the increased rate of climate-related disasters; and increases in global population, living standards and the development of countries. The production of cement and concrete for new buildings has made sand the second most consumed resource after water (The Sandshed 2019).

Mining can be both a threat and an opportunity to achieve SDG 13: a threat because it contributes to the destruction of carbon sinks, biodiversity loss and atmospheric greenhouse gas (GHG) emissions; an opportunity because it furnishes the minerals needed to decarbonize our societies and provide clean energy for all, and it is a hotbed of innovation for decarbonizing industrial processes.

Up until today, the mining sector has been one of the main emitters of GHG emissions (Rüttinger and Sharma 2016). A study with data from 2017 calculated that coal mining, and the production of non-metallic minerals, iron, steel and non-ferrous metals were responsible for approximately 18% of annual global GHG emissions and that a quarter of annual GHG emissions globally came from coal combustion – see Figure 13.1 (ASN Bank, Navigant - A Guidehouse Company December 2019). Data from 2012 show the same trend (Ecofys 2016). These emissions would be much higher if indirect mining contributions (i.e. deforestation, cement production, construction) were also taken into account. For gold alone, global gold industry GHG emissions in 2018 were estimated to be 126 million tonnes of $CO_2$eq (World Gold Council 2019). Although there are similar studies on the impacts of minerals such as iron, steel or coal, in-depth studies for other minerals needed to achieve sustainable futures would help establish mineral-specific emission reduction pathways.

There is also tremendous scope for improvement: compared to other sectors, mining and metals underperforms across most key performance indicators for climate action (Climate Action 100+ 2019). In a survey of 23 diversified miners, steel manufacturers and industrial conglomerates, Climate Action 100+ found that the sector is the weakest in setting board responsibility for climate change (61%) and nearly all (91%) are members of industry associations that lobby against climate policy (Climate Action 100+ 2019) (see Figure 13.1).

Historically, the mining sector is one of the principal contributors to climate change; especially because one of its products, coal, alone contributed to more than 20% of global cumulative GHG emissions from 1988 to 2015 (Riley 2017). Furthermore, mine sites and related infrastructure are increasingly at risk due to climatic events, such as worsening water scarcity, sea-level rise, extreme weather events and related social upheaval. There are strong moral and commercial arguments for mining to embrace a central role in climate action and decarbonization, and so contribute positively to SDG 13. There is a moral argument because mining has been contributing to climate change. There is a commercial argument because new generations care more about this and will start asking for minerals that are not only responsibly sourced, but also sustainably mined.

## Negative Contributions or Threats to Achieving SDG 13

Besides the significant percentage of current and historical GHG emissions that are attributed to some mining and minerals/metals production and consumption activities, mining activities can also adversely affect the environment and pose a threat to SDG 13 through:

- *Deforestation* – Mining is the fourth highest driver of deforestation and mining in forests is growing (The World Bank 2019b). Currently, 7% of all mines occurring in forested areas are based in tropical rainforest biomes, which have the highest biodiversity values, hosting 50% to 75% of the

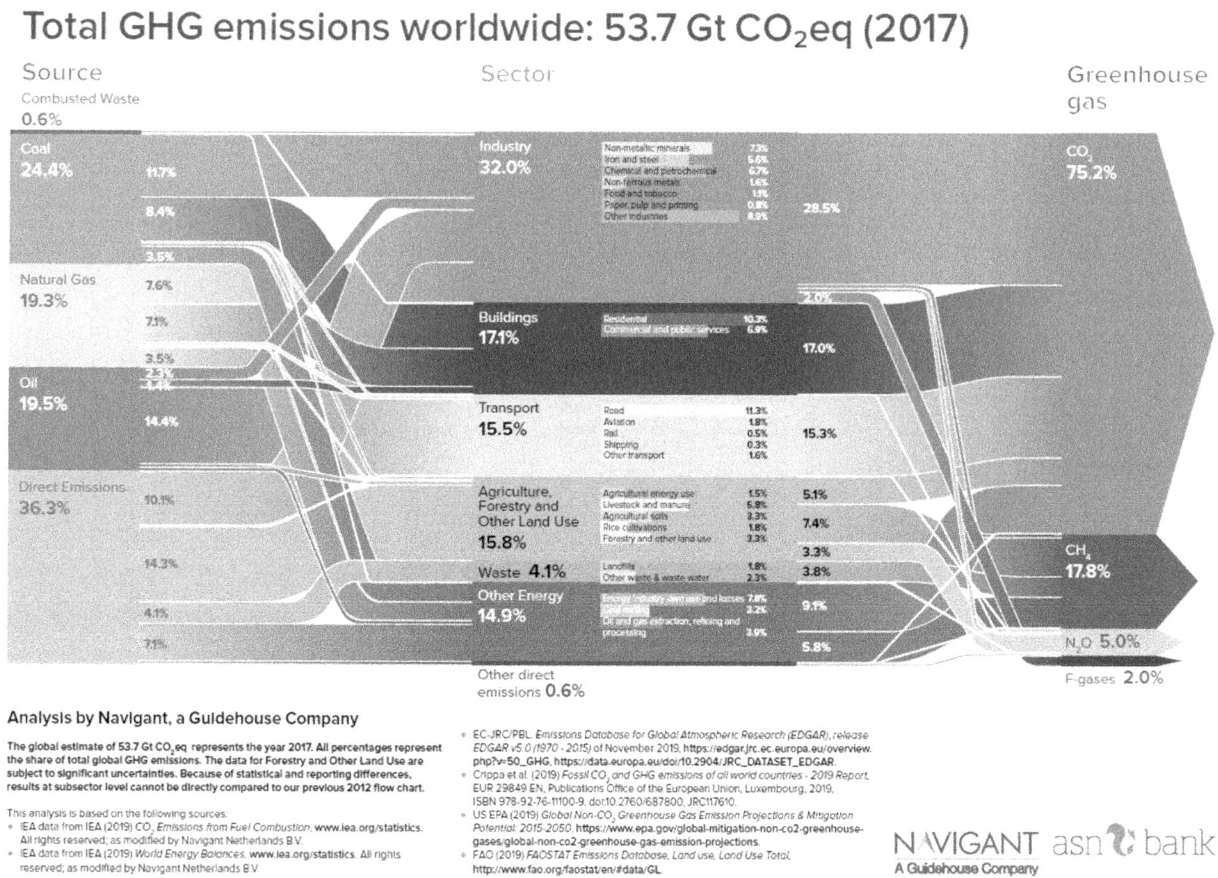


**FIGURE 13.1** Total GHG emissions worldwide chart per source and sector (ASN Bank, Navigant – A Guidehouse Company December 2019).

world's animals and plants, and of course are crucial for carbon storage (The World Bank 2019b). Biodiversity loss is a threat as a result of air, water and soil pollution, and damage to the land.

- *Tailing dam collapses and water sources* – Most recently exemplified by the Brumadinho disaster in Brazil in 2019, where a dam collapse killed 84 people, polluted water sources and caused 125 ha of forest loss, amongst other devastating environmental and social impacts (Associated Press 2019; WWF 2019). Moreover, some mining activities consume a lot of water, which together with climate change, can reduce water availability and create conflict for access to water in water-stressed environments. Water shortages can also pose a threat to the health of forests and therefore they could impact the efficiency of plants' carbon sequestration (Hartmann et al. 2018).
- Causing air pollution – From GHG emissions, dust pollution and the release of toxic elements into the air when refining mineral ores, such as mercury, lead, arsenic or cadmium. Artisanal and small-scale gold mining is the biggest anthropogenic contributor to global mercury emissions (EPA 2018). With the expected increase in temperatures due to climate change, an increased mobilization of legacy mercury reservoirs is expected (USDA 2013), posing a higher threat to ecological systems.
- *Damaging the land* – By physically destroying the land at and beyond the mining site due to the increased influx of people, socio-economic changes and the construction of mining infrastructure such as roads, railways, and ports. Mining and its associated infrastructure frequently adversely impact biodiversity and forests (The World Bank 2019b).
- *Displacement of communities* – In some cases mining activities displace local communities who are often the best guardians of ecosystems, biodiversity and cultural heritage. Conservation outcomes (protection of forests and curbing biodiversity loss) are higher in lands controlled by indigenous communities (Tauli-Corpuz, Alcorn and Molnar 2018). Deforestation rates on customary lands are half of what it is elsewhere (Tauli-Corpuz, Alcorn and Molnar 2018). The displacement of these communities can severely impact the achievement of SDG 13, and mining and extractives are the sectors with the highest environmental defenders being killed in recent years (Brown 2019).
- *Lobbying to intimidate and kill off climate action* – Most recently highlighted in Australia, where the mining industry has a huge influence over the country's climate policies (Knaus 2019). In that case, three of the world's largest mining companies were found to run sophisticated operations lobbying political parties to kill off climate action. Furthermore, the sector has allegedly been working closely with media to avoid any criticism towards them (Knaus 2019).
- *Lack of coordination with other sectors* – Where co-location of mining with other economic activities, such as agriculture or logging, tends to exacerbate negative climate impacts (Ravikumar et al. 2018).
- *Other not fully known impacts* – Deep-sea mining and space mining are considered by many to be vital for a greener future. The impacts of the extraction of minerals in these environments are not fully understood, and could be catastrophic for ecosystems or enormously carbon intensive.

## Positive Contributions or Opportunities to Achieve SDG 13 Key Targets

Although mining is a threat to the achievement of SDG 13, if well managed, mining could not only have a neutral impact, but contribute positively to achieving SDG 13:

- **Target 13.1:** Minerals are needed to *strengthen resilience and adaptive capacity to climate-related hazards and natural disasters*, including in the mining sector which is also vulnerable to climate change. The governments of Fiji and Jamaica have already taken steps to use minerals sourced locally for post-disaster reconstruction activities (IISD 2019). Moreover, certain minerals can enhance natural weathering processes which draw carbon out of the air and sequester it as a newly formed rock (Geoengineering Monitor 2018).
- **Target 13.2:** The increased scrutiny of mining activities by different actors in the mining sector is having a positive impact to *integrate climate change measures into national policies*, strategies and planning. There are many private and public initiatives looking to address the negative impacts of mining. For instance, copper mines in Chile are taking meaningful action to minimize their carbon footprint (Cox 2019) (see Table 13.1).
- **Target 13.3:** *A range of partnerships and initiatives are emerging to improve education, raising awareness and human and institutional capacity on climate mitigation, adaptation, impact reduction and early warning.* For example, the EU Raw Materials Initiative has been crucial to improving understanding in the sector, raising awareness and improving institutional capacity to better manage raw materials in a sustainable way. Within the mining sector specifically, multilateral organizations, such as the World Bank's Climate-Smart Mining Facility and the Intergovernmental Forum on Mining, Minerals, Metals and Sustainable Development (IGF) convene government and private sector events to learn about and inspire climate action in the mining sector (The World Bank 2019; IISD 2019).
- **Target 13.a:** *Implement the UN framework convention on climate change.* Implement the commitment in which developed country parties should mobilise jointly $100 billion annually by 2020 from all sources to address the needs of developing countries in the context of meaningful mitigation actions and transparency.
- Mining can indirectly help achieve the implementation of the UN framework convention by providing minerals to decarbonize the world and provide clean energy to developing countries. The Climate Smart Mining Facility will also act as a vehicle through which the UNFCCC can be implemented.
- **Target 13.b:** *Promote mechanisms to raise capacity for effective climate-change-related planning and management in least developed countries, focusing on women, youth and local and marginalized communities.* Two mining initiatives that can help achieve this target are, (1) the ACP-EU development mineral program that builds the capacity of countries in Africa, the Caribbean and the Pacific to improve the management of development minerals for climate resiliency (UNDP 2019); and (2) the International Women in Mining, a network of women in the mining industry that aims at making the sector more inclusive and diverse (IWIM 2019).

**TABLE 13.1**
**Overview of Initiatives That Are Relevant to Reconciling Mining with SDG 13**

| Initiative | Explanation |
|---|---|
| | **Government regulations and public policies** |
| *Upstream focused* | |
| **In country regulations** | Environmental regulations for mining vary but generally encompass regulations on Environmental and Social Impact Assessments (ESIA), mine closure, tailings management, etc. However, legal or regulatory frameworks on ESIA in most countries do not require a climate change impact assessment (IISD 2019).<br>Canada, an important mining nation, is actively trying to decarbonize mining and exploration, and is playing a strategic role in the green energy transition (IISD 2019). |
| *Operational risk management focused* | |
| **OECD Due Diligence Guidance for Responsible Supply Chains of Minerals from Conflict-Affected and High-Risk Areas** | This guidance provides detailed recommendations to help companies source minerals following the principles of responsible business conduct, so they can avoid complicity in the worst human rights violations and white-collar crimes, and avoid contributing to conflict in their minerals sourcing practices. There are no recommendations on addressing the environmental impacts of sources though businesses have scope to include such issues in their supply chain due diligence activities at their discretion. |
| **Green New Deals (Europe, UK and US)** | The Green New Deals are proposed legislations that aim to address climate change and economic inequality. If we are to achieve a truly green transition, these legislations will need to ensure that minerals used to address climate change are extracted in ways that are climate-smart and follow best environmental practice. |
| | **Industry standards and public–private partnerships** |
| **Initiative for Responsible Mining Assurance (IRMA)** | IRMA defines good practices for what responsible mining activities should look like, including stringent environmental requirements across a range of relevant issues as well as greenhouse gas emissions. IRMA's GHG emissions requirements intend to minimise climate change impacts through reduced energy consumption, increased energy efficiency and reduced GHG emissions. They include: incorporating a GHG policy, quantifying emissions, establish emissions reduction strategies and reporting the policy, the strategies and the GHG emissions. |
| **Responsible Steel** | Responsible Steel is a standard and certification initiative. Its mission is to maximize the contribution of steel to a sustainable society. It includes Principle 8 on Climate Change and Greenhouse Gas Emissions. This Principle requires a corporate commitment to achieve the goals of the Paris Agreement, to disclose climate-related financials, to calculate site-level GHG emissions and intensity, to develop site-level GHG reduction targets and to disclose and report site-level GHG or CO2 emissions. |

(*Continued*)

**TABLE 13.1 (CONTINUED)**
**Overview of Initiatives That Are Relevant to Reconciling Mining with SDG 13**

| Initiative | Explanation |
|---|---|
| **Towards Sustainable Mining (TSM)** | TSM is the Mining Association of Canada's commitment to responsible mining, which includes a protocol and a reference guide on energy and GHG emissions management. The energy use and GHG emissions management protocol consists of three indicators to assess if a facility has implemented a comprehensive system for energy use and GHG emissions management. |
| **Better Coal** | Better Coal is an industry standard working towards a responsible coal supply chain. Member power companies conduct due diligence on their coal suppliers in accordance with ten principles which include a commitment to prevent pollution. Companies are required to quantify significant direct emissions (including GHG emissions) from facilities and indirect emissions to an acceptable degree of accuracy. Companies are also required to develop and implement operating procedures related to minimizing fugitive emissions. |
| **Responsible Jewellery Council (RJC)** | The RJC is an organization that has been established to promote responsible social, ethical and environmental practices in the jewellery sector including along jewellery supply chains up to and including mines. Its Code of Practices includes a requirement to reduce GHGs and pursue materials efficiency (26.2.b), although it does not establish a particular reduction target for GHG emissions. |
| **Aluminium Stewardship Initiative (ASI)** | ASI is a global standard-setting organization that aims to maximize aluminium's contributions to a sustainable society. ASI's requirements include a commitment to reduce GHG emissions from a lifecycle perspective, including the disclosure of GHG emissions and energy use, as well as setting and publishing GHG emissions reduction targets and plan. There are three special requirements only for aluminium smelters, including:<br>1. Putting in place management systems, evaluation procedures and controls to limit direct GHGs<br>2. Scope 1 and 2 GHG emissions from aluminium production are below 8 tonnes Co2eq per metric tonne of aluminium by 2030 (only for aluminium smelters in production up to and including 2020)<br>3. Scope 1 and 2 GHG emissions from the aluminium production are below 8 tonnes Co2eq per metric tonne of aluminium (only for aluminium smelters starting production after 2020) |
| **World Bank Climate Smart Mining Facility** | This is the first-ever fund facility dedicated to ensuring that mining activities are done in a way that avoids, reduces, restores and offsets the mining sector's damage to the environment. The building blocks of climate smart mining are:<br>*Climate change mitigation*: use of renewable energy, innovation in extractive practices, energy efficiency<br>*Climate change adaptation*: forest-smart mining, resource efficiency, innovative tailings solutions |

(*Continued*)

**TABLE 13.1 (CONTINUED)**
**Overview of Initiatives That Are Relevant to Reconciling Mining with SDG 13**

| Initiative | Explanation |
|---|---|
| | *Reducing materials impacts*: adoption of circular economy, recycling of strategic minerals, strategic mineral supply chain management<br>*Creating market opportunities*: de-risking investments, enabling carbon markets, robust geological data |
| **ACP-EU development minerals program** | The ACP-EU development mineral program builds the capacity of countries in Africa, the Caribbean and the Pacific to improve the management of development minerals for climate resilience by improving legal frameworks or developing a nationwide mining register to enable the government to locate post-disaster reconstruction materials. |
| | **Corporate policies and programs** |
| **Mines powered by renewable energy** | There are several examples of mining companies transitioning to renewable energy. The "Renewable Power of Mines" report highlights examples, and bottlenecks impeding progress (Maennling and Toledano 2018). |
| **Monitoring, reporting and disclosing examples** | Many initiatives, such as CDP, Kering's Environmental Profit and Loss tool, the Global Reporting Initiative and suchlike, are helping companies monitor, report and disclose their impacts. |
| **Environmental restoration examples** | Private and civil society-led programs are emerging that seek to reprocess mine tailings to generate the capital necessary to restore the land. RESOLVE's Salmon Gold is reprocessing gold mine tailings in salmon streams in the Yukon, British Columbia and Alaska, for example. |
| **Financing examples** | Innovative finance and emerging technologies, such as biodiversity offsetting schemes, green finance, Environmental Social and Governance (ESG) criteria, amongst others, incentivize and enable mining companies to improve their climate performance. |

Looking beyond mining's contributions to the targets of SDG 13, what should mining activities look like if we are to fully combat climate change in line with SDG 13?

## Vision: Reconciling Mining with Nature and People

The mining sector is facing increasing societal pressure to take action on climate change and align with efforts to build sustainable futures. Fortunately, there are many community initiatives, government regulations, public policies, industry standards, public–private partnerships, corporate policies and private programs of relevance to reconcile mining with SDG 13. Table 13.1 shows a non-exhaustive list of initiatives.

Many of these initiatives have already generated positive impacts, but most also have gaps. First, international *downstream* regulations and industry standards such as the OECD Due Diligence Guidance or the EU Conflict Minerals Regulation do not have an environmental focus. Second, climate action is still predominantly a voluntary prerogative for mining companies, where compulsion to act is driven more by investor, shareholder, market, and civil society pressure, than national legislation.

Third, where national regulations do exist, they vary in terms of scope and quality of enforcement. There is often a wide gap in stringency and the quality of enforcement of environmental regulations between the jurisdictions where mining companies are domiciled and those in which they operate; only in some cases does domestic regulation extend over international activities (e.g. Canada). Regulatory misalignment means only progressive companies compel themselves to go beyond compliance; without adequate board-level oversight, accountability for missing climate targets will be weak.

These gaps highlight that the mining sector has much to do in order to better contribute to the thriving of nature and society, but it is somewhat clear where the sector needs to go. Mines of the future are envisioned to:

- Be net carbon zero by operating with renewable energy and following the mitigation hierarchy, a set of four prioritized steps to alleviate environmental damages: avoid, mitigate or reduce, restore and finally offset.
- Follow the World Bank's Forest Smart Mining Principles (The World Bank 2019b) and the Climate Smart Mining building blocks (The World Bank 2019c).
- Act with other members of society and sectors to push for a just transition.
- Adopt circularity into business models, build awareness on where the mining sector fits in the materials economy and shape a more sustainable supply and use of materials. An integrated business model incorporating both raw materials extraction and recycling allows mining companies to find synergies in materials sourcing and resource recovery, to meet environmental quality standards and also position itself for increasing demand for urban mining. Circular economy strategies in only five areas (cement, aluminium, steel, plastics and food) could reduce the carbon emissions from goods production by almost half (Ellen Macarthur Foundation 2019).

Therefore, mining companies must lead the way, prioritizing climate-smart approaches that will repair for historical damage, support the global transition to a green economy, protect human rights and ultimately also protect and build value for their shareholders and society at large (Table 13.2). It is incumbent on mining investors to release capital for investments that support the introduction of climate-smart mining practices, to enforce investment terms that prioritize climate-smart mining, and to establish mechanisms for oversight and reporting that ensure their mining investments support rather than obstruct climate action. If they do not, they risk undermining the value of their investments as the mining sector will face increasing challenges by key stakeholders such as regulators, civil society and industry pressure groups who will increasingly see climate action as a crucial component of achieving the social license to operate.

Many governments, companies, communities and civil society organizations are trying to build an environment that is conducive to help mining companies and investors to implement the practices above.

**TABLE 13.2**
**Mining Companies and Their Investors Can Lead the Way on Eight Fronts**

| | |
|---|---|
| Avoidance | Avoid or limit mining in areas of Critical Natural Habitat or where nature-based solutions for climate change adaptation are a priority (e.g. as natural flood defences), especially in the tropics, and privilege (re-)mining brownfield sites over greenfield sites. |
| Circularity | Build circularity into industrial processes and product design, prioritizing materials efficiency and adopting an industrial ecology approach. |
| Consistent Policy | Ensure that their climate change policy is consistent with the positions taken by industry associations of which they are a member. |
| Decarbonization | Decarbonize the mining sector's products and operations. |
| Diverse Solutions | Support and invest in financial, technological and natural solutions for carbon sequestration through all phases of the mining lifecycle, including legacy sites. |
| Restoration | Restore post-mining landscapes to benefit natural carbon storage and other ecological values. |
| Repair | Support and invest in climate repair to compensate for historical damage and benefit natural carbon storage. |
| Sustainable Production | Expand the sustainable production of green economy minerals for a sustainable future. |

## LESSONS LEARNED

The main lessons learned are:

- The consumption of minerals has increased exponentially over the past century, as have GHG emissions, and will continue to do so, as recycling alone is currently not enough to meet demand, and for certain metals may never be.
- The mining sector can be a threat as well as an opportunity to achieve SDG 13:
  - Threats: deforestation, tailing dam collapses, air pollution, land damages, displacement of communities and lobbying to intimidate and kill off climate action.
  - Opportunities: strengthen resilience and adaptive capacity to climate-related hazards and natural disasters, integrating climate change measures into national policies, act as a vehicle through which the UNFCCC can be implemented (via the Climate Smart Mining Facility).
- The mining and minerals sector has a responsibility to achieve SDG 13 due to its contributions to climate change, its unique position in helping address it and its vulnerability to it.
- Although there are gaps encountered in the different initiatives we do know where the sector needs to go. Mines of the future are envisioned to:
  - Be net carbon zero
  - Follow the mitigation hierarchy

    - Follow the forest-smart mining and climate-smart mining principles
    - Act with other members of society and sectors to push for a just transition
    - Adapt to a circular economy
- Therefore, mining companies and their investors should lead the way on eight fronts:
    - Avoid or limit mining in areas of Critical Natural Habitat
    - Decarbonize the mining sector
    - Build circularity and industrial ecology into industrial processes
    - Expand sustainable production of green economy minerals
    - Restore post-mining landscapes
    - Invest in climate repair
    - Invest in and share diverse technologies for carbon sequestration
    - Ensure that their climate change policy is consistent with the positions taken by industry associations of which they are a member

## ACKNOWLEDGEMENT

Kirsten Hund and Daniele La Porta.

## REFERENCES

ASN Bank, Navigant - A Guidehouse Company. 2019, December. Total GHG Emissions Worldwide: 53.7 Gt CO2eq (2017).

Associated Press. 2019, January 30. The Weather Channel. Accessed November 15, 2019. https://weather.com/science/environment/news/2019-01-30-brazil-dam-collapse-river-pollution.

BHP. 2019, July 23. BHP to Invest US$400m to Address Climate Change. Accessed November 22, 2019. https://www.bhp.com/media-and-insights/news-releases/2019/07/bhp-to-invest-us400m-to-address-climate-change.

Brown, Alleen. 2019, July 30. More than 160 Environmental Defenders Were Killed in 2018, and Many Others Labeled Terrorists and Criminals. Accessed November 24, 2019. https://theintercept.com/2019/07/30/criminalization-environmental-activists-global-witness-report/.

Climate Action 100+. 2019. Climate Action 100+ –2019 Progress Report.

Cox, Lisa. 2019, October 22. BHP Plans to Replace Coal with Renewables at Two Huge Copper Mines in Chile. *The Guardian*. https://www.theguardian.com/business/2019/oct/22/bhp-replace-coal-renewables-two-huge-copper-mines-chile.

Ecofys. 2016, May. World GHG Emissions Flow Chart. *Eldis*. Accessed November 15, 2019. https://www.eldis.org/document/A100143.

Ellen MacArthur Foundation. 2019. *Completing the Picture: How the Circular Economy Tackles Climate Change*. Ellen MacArthur Foundation.

EPA. 2018, November 21. United States Environmental Protection Agency. Accessed November 15, 2019. https://www.epa.gov/international-cooperation/mercury-emissions-global-context.

Geoengineering Monitor. 2018, May 9. Geoengineering Monitor. Accessed November 26, 2019. http://www.geoengineeringmonitor.org/2018/05/enhanced-weathering-factsheet/.

Hartmann, Henrik, Catarina F. Moura, William R.L. Anderegg, Nadine K. Ruehr, Yann Salmon, Craig D. Allen, Stefan K. Arndt, et al. 2018. Research Frontiers for Improving Our Understanding of Drought-Induced Tree and Forest Mortality. *The New Phytologist*.

Hogan, Michael. 2019, November 1. Anti-Plastic Trend Raising Aluminium Demand. Accessed November 15, 2019. https://uk.reuters.com/article/us-metals-lmeweek-novelis-lmeweek/anti-plastic-trend-raising-aluminium-demand-novelis-idUKKBN1XB46I.

IISD. 2019. *IGF Bulletin – 15th AGM of the International Forum on Mining, Metals, Minerals and Sustainable Development.* Geneva: IISD.

IWIM. 2019. International Women in Mining. Accessed November 15, 2019. https://internationalwim.org/.

Jamasmie, Cecilia. 2018, June 25. Lithium Demand from Battery Makers to Almost Double by 2027. Accessed November 15, 2019. https://www.mining.com/lithium-demand-battery-makers-almost-double-2027/.

Knaus, Christopher. 2019, October 10. Mining Firms Worked to Kill off Climate Action in Australia, Says Ex-PM. *The Guardian.*

Maennling, Nicolas, and Perrine Toledano. 2018. *The Renewable Power of the Mine.* Columbia Center on Sustainable Investment.

Ravikumar, Ashwin, Anne M. Larson, Rodd Myers, and Tim Trench. 2018. Inter-Sectoral and Multilevel Coordination Alone Do Not Reduce Deforestation and Advance Environmental Justice: Why Bold Contestation Works When Collaboration Fails. *Environment and Planning C: Politics and Space* 36(8) 1437–1457.

Riley, Tess. 2017, July 10. Just 100 Companies Responsible for 71% of Global Emissions, Study Says. *The Guardian.*

Rio Tinto. n.d. Energy and Climate Change. Accessed November 15, 2019. https://www.riotinto.com/ourcommitment/energy-and-climate-change-24291.aspx.

Rüttinger, L., and V. Sharma. 2016. Climate Change and Mining - A Foreign Policy Perspective. *Climate Diplomacy.*

Tauli-Corpuz, Victoria, Janis Alcorn, and Augusta Molnar. 2018. *Cornered by Protected Areas.* Right and Resources Initiative.

The Sandshed. 2019. The Global Rise of Illegal Sand Mining. Accessed November 15, 2019. http://arcg.is/040Lav.

The World Bank. 2017. *The Growing Role of Minerals and Metals for a Low Carbon Future.* Washington, DC: Wolrd Bank Publications.

The World Bank. 2019a. Climate-Smart Mining: Minerals for Climate Action. Accessed November 25, 2019. https://www.worldbank.org/en/topic/extractiveindustries/brief/climate-smart-mining-minerals-for-climate-action.

The World Bank. 2019b. *Mining: Large-Scale Mining on Forests (LSM).* Washington, DC: The World Bank.

The World Bank. 2019c. New World Bank Fund to Support Climate-Smart Mining for Energy Transition. May 1. Accessed November 15, 2019. https://www.worldbank.org/en/news/press-release/2019/05/01/new-world-bank-fund-to-support-climate-smart-mining-for-energy-transition.

UNDP. 2019. ACP-EU Development Minerals Programme. Accessed November 15, 2019. http://developmentminerals.org/index.php/en/.

USDA. 2013, February 20. Mercury Impacts to the Environment. Accessed November 22, 2019. https://webcam.srs.fs.fed.us/impacts/mercury/.

Vidal, O., C. Rostom, C. Francois, and G. Giraud. 2017. Global Trends in Metal Consumption and Supply: The Raw Material-Energy Nexus. *Elements* 13(5) 319–324.

World Gold Council. 2019. Gold and Climate Change: Current and Future Impacts. October 23. Accessed November 15, 2019. https://www.gold.org/goldhub/research/gold-and-climate-change-current-and-future-impacts.

WWF. 2019. Dam Burst in Brumadinho Caused 125 Hectares of Forest Loss. January 29. Accessed November 25, 2019. https://www.wwf.org.br/informacoes/english/?uNewsID=69542.

GOAL 14

# Conserve and Sustainably Use the Oceans, Seas and Marine Resources for Sustainable Development

*Samantha Smith*

## CONTENTS



## SUMMARY TABLE

| | |
|---|---|
| Key stakeholders | • International Seabed Authority (ISA), member states, sponsoring states (states that sponsor contractors), ISA observers<br>• ISA contractors (entities with contracts to explore for seafloor minerals in the area beyond national jurisdiction)<br>• Coastal States (where applicable)<br>• National and regional governments (where applicable)<br>• Local communities (where applicable)<br>• Other exploration/mining companies, processors and traders, and international bodies (e.g. IMO)<br>• NGOs, academic/scientific community |
| Direct mining contribution/drivers | Main drivers are growth-in-demand, related to:<br>• Population growth<br>• Urbanisation<br>• Low-carbon (clean energy) technologies |

| | |
|---|---|
| Contribution to SDG 14 | Negative contributions:<br>• Habitat and biodiversity loss, related to mining and sediment plumes<br>• Potential threat to marine food chains and the coastal communities that rely on them<br>• GHG emissions from fossil-fuel–based power for exploration, and transportation, ultimately contributing to ocean acidification and sea-level rise<br>Positive contributions:<br>• Sustainable use of oceans through international law as reflected in the United Nations Convention on the Law of the Sea (UNCLOS)<br>• Sustainable management and protection of marine and coastal ecosystems (sustainable management through a rigorous environmental baseline and impact assessment process, the development of environmental management and monitoring plans, establishment of regional environmental management plans, etc.)<br>• Supporting the conservation of at least 10 to 30 per cent of coastal and marine areas, based on best available scientific evidence<br>• Benefits to Small Island Developing States<br>• Marine scientific research, significant contributions to scientific knowledge about the marine environment<br>• Research capacity development<br>• Provision of minerals for low-carbon technologies, reducing $CO_2$ emissions and ocean acidification<br>• Transferability to other ocean sectors<br>• Link between ocean health and mining provides an entry point for a larger debate on socio-environmental standards |
| Key regulations and public policies | Examples include:<br>• United Nations Convention on the Law of the Sea (UNCLOS)<br>• ISA's Mining Code (exploration regulations, [draft] exploitation regulations, guidelines, policies)<br>• ISA Regional Environmental Management Plans<br>• ISA Recommendations for Environmental Baseline Studies<br>• In-country regulations (where applicable)<br>• IFC Performance Standards on Environmental and Social Sustainability<br>• Equator Principles (as applicable)<br>• International Convention for the Prevention of Pollution from Ships (MARPOL)<br>• London Convention |
| Corporate policies and private programmes (and others) | Examples include:<br>• IMMS Code for Environmental Management of Marine Mining<br>• UN Global Compact Sustainable Ocean Business Action Platform Guidance Documents (in prep)<br>• EU-SPC Deep Sea Minerals Project<br>• The World Bank Climate Smart Mining Facility<br>• ICMM Sustainable Development Principles<br>• Initiative for Responsible Mining Assurance (IRMA) |

## ANALYSIS OF THE MARINE MINERALS SECTOR'S CONTRIBUTION TO SDG 14

Covering over 70 per cent of our planet's surface, the oceans have always been critical to the well-being, prosperity and interconnectivity of humanity. This reliance has never been greater than it is today; the realities of population growth and an unprecedented increase in urbanisation are putting enormous pressure on already strained resources, systems and infrastructure. These are extremely complex and pressing challenges and ones for which there is no simple solution.

One certainty is that the necessary and urgent investment in sustainable infrastructure and transport, renewable energy generation, and clean energy storage is vastly increasing demand for specific metals. While in the long term we may be able to achieve a circular economy in metals, recycling, or 'urban mining', alone will not give us the prodigious quantities of metals we need in the face of extraordinary growing demand (The World Bank, 2020).

This is acknowledged by numerous internationally recognised bodies. For example, the World Bank forecasts that, in a scenario where climate change is limited to +2°C, production of graphite, lithium and cobalt will need to be significantly ramped up by more than 450 percent by 2050 – from 2018 levels – to meet demand from energy storage technologies. These projections do not include the associated infrastructure needed to support the deployment of these technologies (such as transmission lines) or the physical parts (such as the chassis of newly built electric vehicles) (The World Bank 2020).

Similarly, the Institute for Sustainable Futures (ISF) calculates that in a scenario where global temperature rise is limited to less than 1.5 degrees, demand for cobalt will require 423 per cent of known reserves by 2050, for nickel it will be 136 per cent and for lithium 280 per cent (Dominish et al., 2019). Again, this only represents demand from renewable energy and storage.

In addition to addressing climate change, we need to support a rapidly increasing global population and unprecedented urbanisation. The UN estimates that by 2100 there will be almost 11 billion people on the planet, with 68 per cent of that population living in urban areas (United Nations, 2018 and 2019). Urban infrastructure is metal intensive, and we need to accommodate another 3.1 billion city-dwellers over the next 80 years. This means that we need to build the equivalent of another 367 cities the size of New York between now and the end of the century; that's an average of one New York City every 11 weeks. How we resource this in a sustainable manner is one of the most critical challenges facing humankind today.

One solution can be found in the deep ocean. There is a large quantity of potato-sized rocks on the seafloor in the Clarion Clipperton Zone (CCZ) in the North Pacific Ocean between Mexico and Hawaii that contains the exact metals we need for crucial technologies such as electric vehicle batteries, wind turbines and solar panels. They also contain vast quantities of copper, one of the most versatile metals that is used in a range of consumer, construction and industrial applications.

The seabed minerals industry actively welcomes strict regulation that enforces environmental responsibility, transparency and conservation. With successful and ongoing partnerships between governments, communities, civil society, international scientists and academics, and responsible contractors, the seabed mineral industry aims to make positive contributions to all Sustainable Development Goals (SDGs). Specifically, there is clear alignment between the industry and SDG 14, as well as 9, 11, 12 and 13.

SDG 14 (Life Below Water) relates to the conservation and sustainable use of our oceans, seas and marine resources. Seabed minerals exist in three forms: as polymetallic nodules, polymetallic sulphides and cobalt-rich crusts. This chapter focuses on polymetallic nodules, potato-sized rocks that lie on the seabed, which in the CCZ is 4.5 km below the surface. Polymetallic nodules, as their name suggests, are multimetal and it is estimated that the nodules of the CCZ alone boast 1.2 times more manganese, 1.8 times more nickel and 3.4 times more cobalt than all known land-based reserves combined (Hein et al., 2013). These are some of the key metals we need to help mitigate climate change. The nodules also contain significant amounts of copper, molybdenum, tungsten, lithium, niobium, and rare earth oxides (REOs).

While responsible, existing land-based mines must remain part of the solution, opening new mines in increasingly remote, ecologically sensitive locations could come with significant environmental and social costs, such as deforestation, waste due to overburden and tailings, and displacement of communities, among others.

As high-grade resources are depleted, the gross energy requirement of extracting terrestrial sources increases (e.g. Elshkaki et al., 2016). The mining industry has historically been one of the most energy-intensive industrial sectors, and hence one of the largest contributors to global $CO_2$ emissions. A single CCZ deposit contains as much metal as three separate mines on land, and no digging is required, resulting in much lower energy requirements to extract the same amount of metal. Less $CO_2$ means less ocean acidification and less sea-level rise and this, along with providing the metals needed for clean energy infrastructure and to reach circular economy goals, is one of the greatest contributions the marine minerals industry can make to the health of our oceans and the planet.

In the area beyond national jurisdiction, marine minerals exploration and mining are regulated by the UN's International Seabed Authority (ISA). The ISA is an autonomous international organisation established under the 1982 United Nations Convention on the Law of the Sea (UNCLOS) and the 1994 Implementing Agreement relating to the Implementation of Part XI of UNCLOS to administer the mineral resources of the international seabed area (the Area) while also ensuring the effective protection of the marine environment.

The ISA's Mining Code refers to the comprehensive set of rules, regulations and procedures issued by the ISA to regulate prospecting, exploration and exploitation (mining) of marine minerals in the international seabed area. No exploitation can take place without thorough and transparent environmental baseline studies and environmental impact statements assessing the different risks and opportunities. The precautionary approach embedded in the legislative framework has resulted in entities with exploration contracts with the ISA (ISA Contractors) taking a progressive and cautious step-by-step approach to project development to mitigate environmental and economic risk.

The legal framework developed by the ISA for the conservation and sustainable use of the oceans has involved global experts from all relevant disciplines. It is arguably more detailed and robust than any set of regulations that apply to other offshore sectors and has seen international collaboration of world-renowned scientists with environmental, marine and geological expertise.

Moreover, through ongoing collaboration with the independent scientific community, this nascent industry is significantly contributing to marine scientific knowledge and is developing science-based management plans, including strategies to minimise the extent and duration of any impacts.

The industry acknowledges concerns with regard to deep-sea ecosystems, sediment plumes, carbon sequestration and any other potential impact of seabed mineral activities. It is imperative and, as such, accepted by contractors that all of these concerns will need to have been addressed, managed and mitigated to an acceptable level before any commercial activities commence. Such a precautionary approach is written into the Mining Code.

How the marine minerals sector specifically contributes, and can further contribute, to SDG 14 is outlined below:

- **Target 14.1:** The marine minerals sector can prevent and significantly reduce marine pollution, through a focus on high-grade and multi-metal deposits which reduce waste and also through thorough Waste Prevention and Waste Management Plans and adherence to and exceedance of international conventions aimed at minimising waste and marine pollution such the International Convention for the Prevention of Pollution from Ships (MARPOL) and the London Convention on the Prevention of Marine Pollution by Dumping of Wastes and Other Matter (London Convention).
- **Target 14.2:** The marine minerals sector can sustainably manage and protect marine and coastal ecosystems through the use of ecosystem-based approaches for environmental management, and by taking a precautionary approach to project and industry development. The ISA Mining Code requires detailed environmental baseline studies, environmental impact assessments, science-based environmental management and monitoring plans, and closure plans prior to approval of a plan of work for mining. The ISA has also established a Regional Environmental Management Plan (REMP) for the CCZ nodule province in the Pacific Ocean, where current commercial interest is greatest, and is in the process of establishing further REMPs for other areas of potential mining interest. A key aspect of the REMPs is the establishment of Areas of Particular Environmental Interest (APEIs), a network of large conservation areas that must remain protected from the effects of mining. The purpose of the REMPs and APEIs is to protect biodiversity and ecosystem health and function on a regional level, thereby playing an important role to ensure the effective protection of the marine environment. ISA Contractors are also obligated to establish Preservation Reference Zone(s) as an integral part of their mine plans.
- **Target 14.3:** The marine minerals sector can minimise and address the impacts of ocean acidification, including through enhanced scientific cooperation at all levels through appropriate studies during the baseline study and impact assessment phases, including studies on marine acidity (pH), and combining/submitting results into a global database and, through

partnerships with the scientific community, providing information through publication in scientific journals. The mining sector in general can also contribute to this target through the use of low-carbon technologies, resulting in lowered GHG emissions.

- **Target 14.4:** While this target relates to the effective regulation of harvesting and the end of overfishing, the marine minerals sector may be able to contribute to this target through scientific data collection during offshore research campaigns and reporting.
- **Target 14.5:** The marine minerals sector supports the conservation of at least 10 per cent of coastal and marine areas, consistent with national and international law and based on the best available scientific information. As mentioned above, the ISA has established, and is establishing, Regional Environmental Management Plans (REMPs) for areas of potential marine mining interest. A key aspect of the CCZ REMP is the establishment of Areas of Particular Environmental Interest (APEIs), a network of large (400 km × 400 km) conservation areas that must remain protected from the effects of mining. Key purposes of the REMPs and APEIs include the protection of biodiversity and ecosystem health and function on a regional scale. The CCZ APEIs cover 1.44 million km$^2$, which is more than 30 per cent of the CCZ environmental management area.
- **Target 14.6:** This target deals with fishing subsidies and is not applicable to the marine minerals sector.
- **Target 14.7:** The marine minerals sector can contribute to the economic benefits to Small Island Developing States and least-developed countries from the sustainable use of marine resources in a number of ways. The ISA Mining Code seeks to ensure that research, exploitation and exploration of the seabed benefits all humanity, not just technologically advanced countries. It will do this by requiring a portion of the financial benefits from seabed mining to be paid to the ISA, who will then share the proceeds with developing countries through equitable sharing criteria (Lodge, 2019).

Some Small Island Developing States have chosen to become ISA Contractors, through partnerships and/or agreements with commercial entities. This has allowed access to development opportunities that may not have existed otherwise.

Seabed minerals are also vital in the production of clean energy and other cutting-edge technology and innovation, which will be central to the continued and sustainable industrial and economic growth of least developed countries (ISA, 2014).

- **Target 14.a:** The marine minerals sector can increase scientific knowledge, through industry–academic partnerships and publications in the scientific literature. Seabed mineral activities enable scientific and environmental communities to benefit from improved access to marine samples, open up access to data and international research, and enable greater (international) collaboration, which improves our understanding and informs decisions relating to SDG 14.

Many of the deep ocean species that are being identified today would have remained a mystery were it not for seabed mineral exploration. Exploration of deep-sea ecosystems can also reveal new marine genetic resources (MGR) that might lead to the development of new antibiotics and drugs to treat cancer, Alzheimer's and cystic fibrosis, among others. The fledgeling MGR industry can benefit from access to samples collected during seabed mineral exploration.

Deep-sea exploration is driving incredible advances in technology and innovation. In the same way scientific breakthroughs developed for space exploration have become ubiquitous in our daily lives – from mobile phone cameras and scratch-resistant lenses, to water-filtration systems – underwater activities are contributing to innovations in robotic vehicles, remote instrumentation, web services, computational data analysis and artificial intelligence.

The marine minerals sector can also contribute to Target 14.a by developing research capacity through the ISA Training Program (including its Endowment Fund and internship opportunities) and Contractor Training Programs, often aimed at marine scientific research training.

- **Target 14.b:** This target deals with providing access for small-scale artisanal fishers to marine resources and markets and is not applicable to the marine minerals sector.
- **Target 14.c:** This target is aimed at enhancing the conservation and sustainable use of oceans and their resources by implementing international law as reflected in UNCLOS. In the area beyond national jurisdiction, seabed mineral activity is governed through UNCLOS, which provides the legal framework for the conservation and sustainable use of oceans and their resources.

## Lessons Learned

The main lessons learned are:

- The consumption of minerals has increased exponentially over the past century and continues to rise.
- Clean energy technologies are metal intensive.
- Vast amounts of metal reside on the seafloor.
- Our oceans are critical to a healthy planet and ocean resources need to be managed appropriately.
- Taking a precautionary approach to manage our ocean resources, including the requirement for prior Environmental Impact Assessments (EIAs), Environmental Management and Monitoring Plans (EMMPs) and/or Regional Environmental Management Plans (REMPs), is of critical importance.
- REMPs with large protected areas included are key to responsible ocean management.

- High-grade, multi-metal deposits and efficient processing solutions need to be the focus to ensure the mining industry produces significantly less waste going forward.
- The marine mining sector has an incredible opportunity to substantially contribute to achieving SDG 14.
- Industry–academic (and other stakeholder) partnerships can make significant contributions to scientific knowledge and are important for informed, science-based decision-making.
- Co-ordination with other sectors, particularly other maritime sectors, is important.
- There is increasing recognition that mitigating negative impacts on ocean health is imperative. Going the next step and 'giving back' to ensure ocean health should be considered as a shared goal for all (ocean) industries.
- Improvements can be made to capacity building, helping to ensure more developing nations can participate and benefit from marine mineral resources.
- There are many initiatives attempting to reconcile mining with nature; the seafloor mining sector should learn from these and where possible improve upon them.

Marine mineral exploration entities and their investors can contribute in 12 concrete ways:

1. Avoid seafloor mining in, or negatively impacting, Marine Protected Areas;
2. Share data, and contribute to marine scientific knowledge, including for the establishment of further protected areas;
3. Establish and implement regional planning for environmental management; establish Regional Environmental Management Plans, including networks of protected areas that must remain unimpacted by mining;
4. Collaborate with other contractors and other sectors – to share data and improve regional understanding of the environment, including biological connectivity;
5. Contribute to long-term observations of ocean acidification (pH) and other important ocean health parameters;
6. Consider what can be done to ensure the marine minerals sector has a net positive impact on fisheries;
7. Decarbonise the mining sector;
8. Focus on high-grade, multi-metal deposits with an aim of significantly less mining waste;
9. Commit to zero deforestation; no trees need to be removed during the mining process, but entities can ensure that land-based processing occurs in brownfield sites where tree removal also is not needed;
10. Expand capacity-building initiatives with a particular focus on marine scientific research and technology;
11. Restore post-mining seascapes where reasonably practical and economically and technologically feasible;
12. Invest in ocean health initiatives.

## REFERENCES

Dominish, E., N. Florin, and S. Teske. 2019. *Responsible Minerals Sourcing for Renewable Energy*. Report Prepared for Earthworks by the Institute of Sustainable Futures University of Technology Sydney.

Elshkaki, A., T.E. Graedel, L. Ciacci, and B.K. Reck. 2016. Copper Demand, Supply and Associated Energy Use to 2050. *Global Environmental Change* 39: 305–315.

Hein, J.R., K. Mizell, A. Koschinsky, and T.A. Conrad. 2013. Deep-Ocean Mineral Deposits as a Source of Critical Metals for High- and Green- Technology Applications: Comparison with Land-Based Resources. *Ore Geology Reviews* 51: 1–14.

International Seabed Authority. 2014. Deep Sea Mining Environmental, Legal and Technical Challenges for Developing Countries. *ISA Briefing Paper 01/2014*. Accessed 20 December 2019. https://ran-s3.s3.amazonaws.com/isa.org.jm/s3fs-public/files/documents/orem-bp1-2014.pdf.

Lodge, M. 2019. Regulating Deep Sea Mining. *The Economist Group World Ocean Initiative (Blog)*. Accessed 20 December 2019. https://www.woi.economist.com/regulating-deep-sea-mining/.

The World Bank. 2020. *Minerals for Climate Action: The Mineral Intensity of the Clean Energy Transition*. Washington, DC: World Bank Publications.

United Nations Department of Economic and Social Affairs. 2018. 2018 Revision of World Urbanization Prospects. Accessed 20 December 2019. https://population.un.org/wup/.

United Nations Department of Economic and Social Affairs. 2019. World Population Prospects 2019: Highlights. Accessed 20 December 2019. https://population.un.org/wpp/.

## BIBLIOGRAPHY

Clark, M.R., H.L. Rouse, G. Lamarche, J.I. Ellis, Christopher Wayne Hickey, and National Institute of Water and Atmospheric Research (N.Z.). 2017. Preparation of Environmental Impact Assessments: General Guidelines for Offshore Mining and Drilling with Particular Reference to New Zealand. (New Zealand) National Institute of Water and Atmospheric Research.

International Marine Minerals Society. 2011. Code for Environmental Management of Marine Mining. Accessed 20 December 2019. https://www.immsoc.org/IMMS_downloads/2011_SEPT_16_IMMS_Code.pdf.

International Maritime Organization. 2006. 1996 Protocol to the Convention on the Prevention of Marine Pollution by Dumping of Wastes and Other Matter, 1972, as Amended in 2006.

International Seabed Authority. 2012a. Decision of the Council Relating to an Environmental Management Plan for the Clarion-Clipperton Zone. (ISBA/18/C/22) 26 July 2012.

International Seabed Authority. 2012b. ISA Technical Study #10: Environmental Management Needs for Exploration and Exploitation of Deep-Sea Minerals.

International Seabed Authority. 2019. Draft Regulations on Exploitation of Mineral Resources in the Area. (ISBA/25/C/WP.1).

International Seabed Authority, Legal and Technical Commission. 2019. Recommendations for the Guidance of Contractors for the Assessment of Possible Environmental Impacts Arising from Exploration for Marine Minerals in the Area. (ISBA/25/LTC/6).

Le, Jennifer T., Lisa A. Levin, and Richard T. Carson. 2017. Incorporating Ecosystem Services into Environmental Management of Deep-Seabed Mining. *Deep Sea Research II* 137: 486–503.

The PEW Charitable Trusts. 2019. Regional Environmental Management Plans Are Key to Deep-Sea Conservation. Accessed 20 December 2019. https://www.pewtrusts.org/en/research-and-analysis/fact-sheets/2019/07/regional-environmental-management-plans-are-key-to-deep-sea-conservation.

The World Bank. 2019. New World Bank Fund to Support Climate-Smart Mining for Energy Transition. May 1. Accessed November 15 2019. https://www.worldbank.org/en/news/press-release/2019/05/01/new-world-bank-fund-to-support-climate-smart-mining-for-energy-transition.

Vidal, O., C. Rostom, C. Francois, and G. Giraud. 2017. Global Trends in Metal Consumption and Supply: The Raw Material-Energy Nexus. *Elements* 13(5): 319–324.

GOAL 15

# Mining and Life on Land

*Kirsten Dales and Paul Cordy*

## CONTENTS



## INTRODUCTION

On the surface, there are clear trade-offs between mining and achieving land degradation neutrality (LDN)* for biodiversity conservation. Yet, when informal, poverty-driven mining livelihoods and their occurrence in nearly all terrestrial biomes and important biodiversity areas on earth is considered, trade-offs become decidedly less clear. This chapter assesses SDG 15 (life on land) and evaluates the extent to which informal miners are being 'left behind' in global conservation, sustainable land management, and landscape restoration, reflecting one of the most critical interfaces in mineral resource governance and society today.

Minerals and metals underpin nearly all sectors in the global economy (IRP, 2020), and their supply is critical to achieve the United Nations Sustainable Development Goals (SDGs) and successfully implement the Paris Agreement (Ali et al., 2017). Accordingly, mineral resource governance has been placed at the leading edge of the 2030 Development Agenda, with growing consensus that mining must continue for the foreseeable future (Ali et al., 2017; IRP, 2019, 2020). While mining in to critical ecosystems raises concern, the international community must plan for coming decades with rising demand for minerals, metals, and other common commodities (i.e., sand, clay, stone) (Franks, 2015; Franks, in press). In rural economies, continuous use of land for mining, agriculture, livestock production, and fuelwood harvest can undermine conservation, sustainable land management, and rehabilitation, especially where artisanal and small-scale mining (ASM) low-tech,

* Defined by Parties to the UN Convention to Combat Desertification (UNCCD) LDN is a 'state whereby the amount and quality of land resources, necessary to support ecosystem functions and services and enhance food security, remains stable or increases within specified temporal and spatial scales and ecosystems.' As of 2019, 123 countries had set voluntary targets to achieve LDN, referenced in target SDG 15.3.

labour-intensive mineral processing and extraction (Hilson, 2009, p.1) enables subsistence. This chapter considers how mining at various scales impacts the opportunities to reach SDG 15. There is a greater emphasis on ASM in the chapter since this area of activity poses a challenge for land restoration strategies and governance in remote high biodiversity regions.

## MINING AND THE ENVIRONMENTAL IMPACTS OF INFORMALITY

In rural economies, land is the basis for all livelihoods where informal mining, agriculture, and logging coexist as part of a multifunctional working landscape. Mining is often deeply entrenched in the cultural traditions of ASM communities (Ali, 2018), representing not only an emerging rural economy but a way of life for millions of women, men, boys, and girls worldwide.* While global statistics on miner registration remain limited, recent estimates suggest that approximately 70% of all artisanal miners operate in some degree of informality without the security of a permit, extension services, or other forms of support (IGF, 2017). Due to widespread informality, the majority of mines conduct activities without formal exploration or extraction licences, leading to poor allocation of subsurface mineral rights and insecure land tenure, which recurrently drives conflict (Andrews, Elizalde, and Le Billon (2017). Access to land with known mineral potential is critical for sustainable development of the ASM sector (Davidson, 1993; Hentschel et al., 2003; Hilson and Maponga, 2004), yet shortfalls in good governance and deficits in land allocation undermine global ambitions to protect, restore, and promote the sustainable use of terrestrial ecosystems, sustainably manage forests, combat desertification, reverse land degradation, and halt biodiversity loss. Integrated landscape approaches must optimize land allocation and formalization processes through integrated, multi-sectoral strategies that aim to manage spaces, not people, in under-resourced jurisdictions (Mutemeri et al., 2016) and improve performance across the mine life cycle.

While specific impacts vary with geology and extraction methods used to mine a deposit, the impacts closely linked to common characteristics of ASM found in many countries include, among others, low levels of formalization, inadequate legal frameworks, lack of financing, equipment, skills, and knowledge to improve mining practices. In many ways, the informality of the ASM sector limits its potential for progress, and due to limited capital investment it lacks the resources to conduct formal exploration, mine rehabilitation and closure, or post-closure planning (Figure 15.1). While the magnitude of impacts from industrial mining can be profound, evidenced by catastrophic tailings dam bursts (Fernandes et al., 2016) and legacies of mine abandonment (Hattingh, Williams and Corder, 2019) the capacity of large-scale mining actors to mitigate and redress these impacts far exceeds that of artisanal and small-scale miners.

* According to the World Bank's Delve Platform as of May 15th, 2020 there are 41,489,796 people working in ASM, 70% men and 30% women. The largest ASM population is located in South Asia (15,880,000), followed by Sub-Saharan Africa (13,213,800), East Asia and the Pacific (10,552,100), Latin America and the Caribbean (1,812,396), Middle East and Northern Africa (21,500) and Central Asia / Europe (10,000).

**FIGURE 15.1** Multifunctional mining landscapes. Figure depicts a tropical settlement of the global south where informal mining, agriculture, and timber harvest enable subsistence but lack land or mineral rights. (Original design by Kirsten Dales.)

Mining can transform landscapes with direct and indirect impacts on ecosystems and their biodiversity. Environmental impacts from mining are often cumulative, operating at multiple spatial and temporal scales (Getty, and Morrison-Saunders, 2020). Figure 15.2 shows the differences between large-scale mining (LSM) and ASM activities in the context of land-use and the life cycle of mining projects. Although ASM holds potential to drive and diversify local economies, create jobs and reduce extreme poverty, the sector also poses significant risks to sustainable development through a range of potentially negative environmental, occupational, and social impacts. Ecosystems, inland waters, and biodiversity are impacted *during* prospection, exploration, construction, operation, closure, and post-closure stages of mining. Differences in production scale between ASM and LSM are a product of their typical capital investment and access to finance, which in turn affects the level of mechanization, organization, formality, and ability to mitigate adverse impacts.

Direct impacts or threats are synonymous with anthropogenic impacts and proximate pressures. Threats can be past (historic), ongoing, or likely to occur in the future (IUCN, 2006). Direct threats to biodiversity, for instance, refer predominantly to human-mediated activities or processes that have impacted, are impacting, or may impact the status of a plant, bird, fish, insect, or mammal species of interest (e.g., unsustainable mining, logging, agriculture). While direct drivers tend to influence ecosystem structure and function, indirect drivers are more diffuse, altering one or more *direct* drivers through complex interaction across spatial, temporal, and organizational scales. Such interactions can yield unintended consequences.

For example, mining is an emerging threat to biodiversity as a result of direct and indirect drivers (Phillips, 2001; Chomitz et al., 2007; Villegas et al., 2012; Sonter et al., 2018; Camizuli et al. 2018; IPBES, 2018). Direct drivers occur due to habitat change (i.e., loss, degradation, fragmentation), unsustainable use or overexploitation, introduction of invasive species (especially in island ecosystems), climate change, and chemical pollution. While biodiversity offsets *on* mine concessions have become

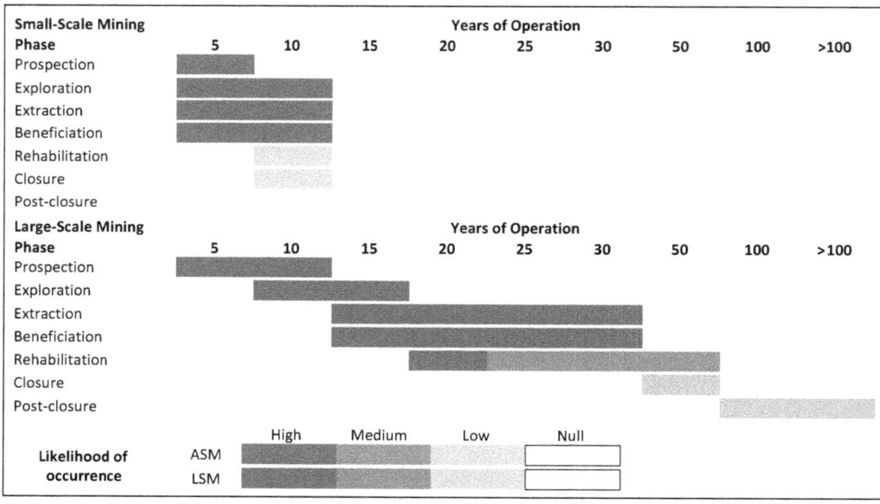


**FIGURE 15.2** Life cycle of small-scale and large-scale mining operations. Figure shows the differences between large-scale mining (LSM) and artisanal and small-scale mining (ASM) activities in the context of land-use and the life cycle of mining projects. Description of categories in figure provided in Table 15.1. (Original design by Kirsten Dales and Bernardo Ranieri.)

a growing focus of LSM operations, habitat destruction and biodiversity losses frequently occur *off* concession due to indirect threats (Phillips, 2001; Sonter et al., 2017). Indirect impacts, such as migration into remote ecosystems, can enhance the threat of deforestation or illegal wildlife poaching. In ASM contexts, mining is rarely the major driver of land-use change leading to deforestation, but indirectly drives change through influxes of miners, traders, and secondary industries to sustain remote operations, encouraging illegal timber and bushmeat harvest (World Bank, 2019), bringing humans into ever closer contact with wildlife. Disturbingly, increased human contact with natural disease reservoirs (i.e., wildlife) and illegal trade are drivers of infectious disease emergence, where proximity to bat species has been traced to Ebola outbreaks in West Africa and corona viruses worldwide (Dominguez et al., 2007; Afelt, Frutos and Devaux, 2018; Valitutto et al., 2020).

The potential for mining to negatively impact the environment varies with commodity, production scale, capital investment, levels of formality, mechanization, and miner education, as well as prevailing political, regulatory, institutional, and physical features of the landscape. However, the global conservation community tends to focus on negative aspects and problems in mining, while too few papers offer pragmatic solutions for the ASM sector. Mineral resources are finite, making the closure of a mining operation inevitable. Planning for closure needs to be a core business practice, and ideally considered in the design phase of an operation – or beginning with the end in mind. (see Figure 15.2; a description of the categories in the Figure are provided in Table 15.1 and 'Summary of Remediation, Reclamation, Rehabilitation, and Restoration Definitions, Targets, Biodiversity Gains, and End-Goals').

**TABLE 15.1**
**Description of Mine Life Stages and the Likelihood of Occurrence between Artisanal, Small-Scale, and Large-Scale Mining Operations**

| Life Cycle Stage | # of Years | ASM | LSM | Description of Life Cycle Stage |
|---|---|---|---|---|
| **Prospecting** | 1–10< | X | X | Mining begins with the search for new deposits. Prospecting occurs in ASM and LSM, both of which often have expert prospectors or 'discovers' of deposits. Due to lack of capital investment, geological data, and land allocation, however, prospecting in ASM is a recurrent and ecologically damaging process. |
| **Exploration** | 1–10 | X | X | Mineral exploration requires significant time, resources, and expertise (e.g., geology, chemistry, geotechnical engineering). Once basic facts about the local geology and economic viability of a deposit are established, feasibility of ore extraction can be assessed and can take up to ten years. Formal exploration rarely occurs in ASM, but in certain contexts ASM-LSM coexistence is helping small-scale miners use geologic data. |
| **Extraction** | 10–30 | X | X | Mineral extraction refers to accessing or removal of minerals from the earth, requiring different mining methods, depending on each type of ore deposit. |
| **Beneficiation** | 10–30 | X | X | In the mining industry or extractive metallurgy, beneficiation is any process that improves (benefits) the economic value of the ore by removing the gangue minerals, which results in a higher-grade product (concentrate) and residual waste streams (tailings, overburden, waste rock, wastewater). |
| **Rehabilitation** | 5–10< | | X | Mine rehabilitation is fundamental to responsible mining. In LSM, mine rehabilitation takes place long before mining commences. Rehabilitation is undertaken progressively during the life of a mine wherever practical and is continually reviewed. Rehabilitation occurs not only at the end of mining operation but is progressive and is ongoing. |
| **Closure** | 10< | | X | Mine closure is the period when mining activities have ceased, and decommissioning and mine reclamation are being completed. Mine closure is generally associated with reduced employment levels, which can have significant negative social impacts as well as more visible environmental impacts. |
| **Post-closure** | 100< | | X | Optimizing safe post-mining land use is key in planning as it sets goals for closure activities. Integrating site-specific rehabilitation plans within overarching regional planning frameworks has become a global best practice. In ASM, mines rarely close, as tailings containing valuable mineral are repeatedly scavenged, discouraging formal closure. |

*Source:* Original design by Kirsten Dales.

## SDG 15 MITIGATION HIERARCHY TO 'AVOID, REDUCE, REVERSE'

Recent shifts in land management policies and practices to achieve 'Land Degradation Neutrality' aim to counterbalance expected losses of productive land resources with the recovery of degraded areas, whereby degradation is avoided, reduced, and reversed through sustainable land management and/or restoration efforts (Figure 15.2). This approach is consistent with the 'Scientific Conceptual Framework for Land Degradation Neutrality' published by the UNCCD Science-Policy Interface (Orr et al., 2017; Cowie et al., 2018), which applies definitions established by the Society for Ecological Restoration: 'Restoration seeks to re-establish the pre-existing biotic integrity, in terms of species composition and community structure, while rehabilitation aims to reinstate ecosystem functionality with a focus on provision of goods and services rather than restoration' (McDonald et al., 2016). See Figure 15.3.

### TRENDS IN PROTECTION OR 'AVOIDING' DEGRADATION

Mining in and around protected areas (PAs) remains a global concern (Phillips, 2001; Villegas et al., 2012; Sonter et al., 2018). Global efforts to protect biodiversity, slow deforestation, and conserve natural habitats rely heavily on the establishment of PAs, land set aside that cannot be deforested or developed. Within the last fifty years, the number and geographic extent of terrestrial and marine protected areas (PAs)* has

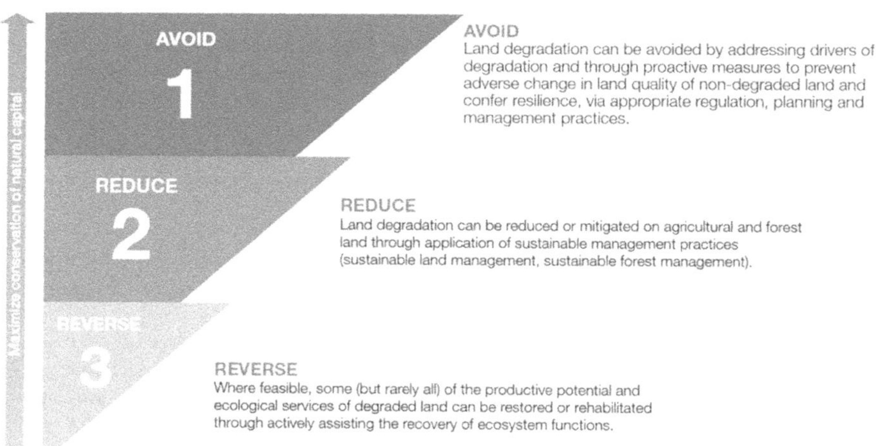


**FIGURE 15.3** Response hierarchy that promotes avoidance over reduction (mitigation) and ultimately reversal (restoration and rehabilitation) of land degradation (*sensu* Orr et al., 2017; Cowie et al., 2018 cited in IRP, 2019).

* According to the 'Protected Planet Digital Report (2020),' there are 244,869 protected areas (PAs) globally, the majority of which are terrestrial, covering 20,359,155 km2 or 15.1% of the earth's land surface. Between 2016 and 2020, terrestrial coverage increased slightly from 14.7% to 15.1%, whereas marine coverage in national waters increased comparatively faster from 10.2% to 18.4%.

consistently increased, demonstrating international progress toward the protection of natural habitats and species. In 2020, 44% of each terrestrial, 41% of each freshwater, and 41% of each mountain Key Biodiversity Area (KBA) were located within PAs, a 12–13% increase since 2000 (ECOSOC, 2020). Globally, 20.3 million km$^2$ of terrestrial ecosystems are PAs, which provide society with many ecosystem services, including carbon storage, a tool for climate mitigation. For instance, intact tropical forests sequester twice as much carbon as planted monocultures (Baccini et al., 2017; Huang et al., 2018).

Unfortunately, informal and illegal miners frequently operate in PAs (Villegas et al., 2012; World Bank 2019) and mineral sourcing from developing countries is rapidly degrading important carbon stores, biologically diverse ecosystems and their benefits. Tropical peatlands (forested wetlands) in Indonesia, the Amazon, and Congo Basin contain vast quantities of irrecoverable carbon where land-use decisions can expand PAs and avoid their degradation or loss (Goldstein et al., 2020). In reality, *de jure* establishment of a PA does not necessarily ensure the protection of species, habitats, or world heritage sites inside its borders. As much of the world's biodiversity lies in countries with high corruption rates, weak property rights, semi-transparent land concessions, instability, or insecurity, the degree to which terrestrial PAs can offer *de facto* protection for carbon (i.e., tropical forests, peatlands) and biodiversity is heavily influenced by the institutional, political, and socio-economic contexts of the countries in which they are created (Abman, 2018). Avoiding land degradation has links to all SDG 15 targets, but especially SDG 15.1 (ecosystem protection), 15.2 (forest conservation), 15.3 (LDN), 15.4 (mountains), 15.5 (biodiversity), 15.7 (wildlife protection), 15.8 (invasive species), 15.9 (integrating biodiversity values into land-use planning) and 15.a.b.c. (finance for nature and climate).

## Sustainably Managing or 'Reducing' Degradation

Land and ecosystem degradation driven by human activities has undermined the well-being of 3.2 billion people worldwide (IPBES, 2018), representing a serious threat to 'life on land' (SDG 15). Mining is a significant global driver of land disturbance, habitat change, and contamination of inland waters and terrestrial ecosystems at multiple levels (de Castro Pena et al., 2017; Seal et al., 2017; Sonter et al., 2018), and cause of some of the highest intensity landscape transformations of all resource sectors (Alvarez-Berros and Mitchell Aide, 2015). Moreover, the impacts of mining can become significant when multiple instances, such as historic rushes or operational scales of mining occur at the same location simultaneously, such as ASM invasion of LSM concessions or *vice versa* (World Bank, 2019). While poorly monitored at global levels, mining is a key driver of human-mediated land degradation (see IPBES, 2018).

According to the UN Secretary General's most recent report on progress made toward the SDGs, the proportion of global forest area fell from 31.9% of total land area in 2000 to 31.2% in 2020, representing a net loss of almost 100 million hectares of forest resources (ECOSOC, 2020). Between 2000 and 2020, forest area increased in Asia, Europe, and North America but significantly decreased in Latin America, sub-Saharan Africa, and South-Eastern Asia, driven by agricultural land conversion (ECOSOC, 2020). Sustainably managing or reducing land-use and land cover change from mining presents unique challenges where informal ASM occurs in regions with already high deforestation rates. Regional changes in forest area between 1990–2015 show greatest losses in Africa and LAC with uneven trends in Asia (see Figure 15.4). Regrettably, efforts to quantify global forest losses have historically lumped mining in with other land uses (World Bank, 2019), potentially mislabelling ASM as agriculture or logging. Consequently, improved data resolution and spatial attribution of ecosystem disturbances due to mining in conservation priorities are needed to understand how, where, and when informal, illegal, or industrial actors are driving permanent land-use change for specific commodity production (Curtis et al., 2018).

Reducing terrestrial ecosystem degradation is relevant to all mining operations, and despite a negative public image, mining is an ancillary driver of deforestation compared to agriculture and forestry, accounting for 5–7% of forest losses annually (Hosonuma et al., 2012; World Bank, 2019). The challenges inherent in reducing the footprints of informal resource sectors notwithstanding, in the case of ASM the situation is further complicated by a complex web of indirect drivers. At the global level, commodity prices, population growth, trade, multilateral commitments, and political will, alongside regional and local factors, including poverty, food insecurity, burden of disease, climate vulnerability, conflict, and presence of armed groups, can undermine sustainable land and forest management efforts. Indeed, combating land degradation and restoring degraded landscapes are urgent global priorities to protect ecosystem structure, function, and the services that enable our life on earth with important SDG 15 targets, especially 15.1 (sustainable land-use in diverse contexts), 15.2 (sustainable forest management), 15.3 (LDN), 15.5 (biodiversity), 15.7 (illegal poaching and illicit trade), 15.8 (invasive species suppression), 15.9 (integrating biodiversity values during mine life and operations), and 15.a.b.c. (mobilizing finance for nature and climate).

## Restoring or 'Reversing' Degradation

According to the world's first comprehensive assessment of land degradation and restoration potential, more than 75% of all land areas on earth have been degraded, with wetlands incurring greatest (area) losses* (IPBES, 2018). Land degradation is eroding nature's contributions to people and biodiversity at alarming rates (IPBES, 2019). Economic benefits of land restoration and rehabilitation projects have been shown to exceed their costs (IRP, 2019; Bullock et al., 2011), where programs to reverse or halt current trends in degradation could generate an estimated USD 1.4 trillion annually (IPBES, 2018). Land restoration can deliver synergetic benefits toward achievement of all SDGs (IRP, 2019), especially SDG 15, alongside the Convention on Biological Diversity (CBD) 2020 Strategic Plan and the Aichi Targets; the Paris Agreement on Climate Change; the UN Strategic Plan for Forests (2017–2030), the UN Decade of

---

* IPBES estimates show global wetlands have been reduced by at least 85%.

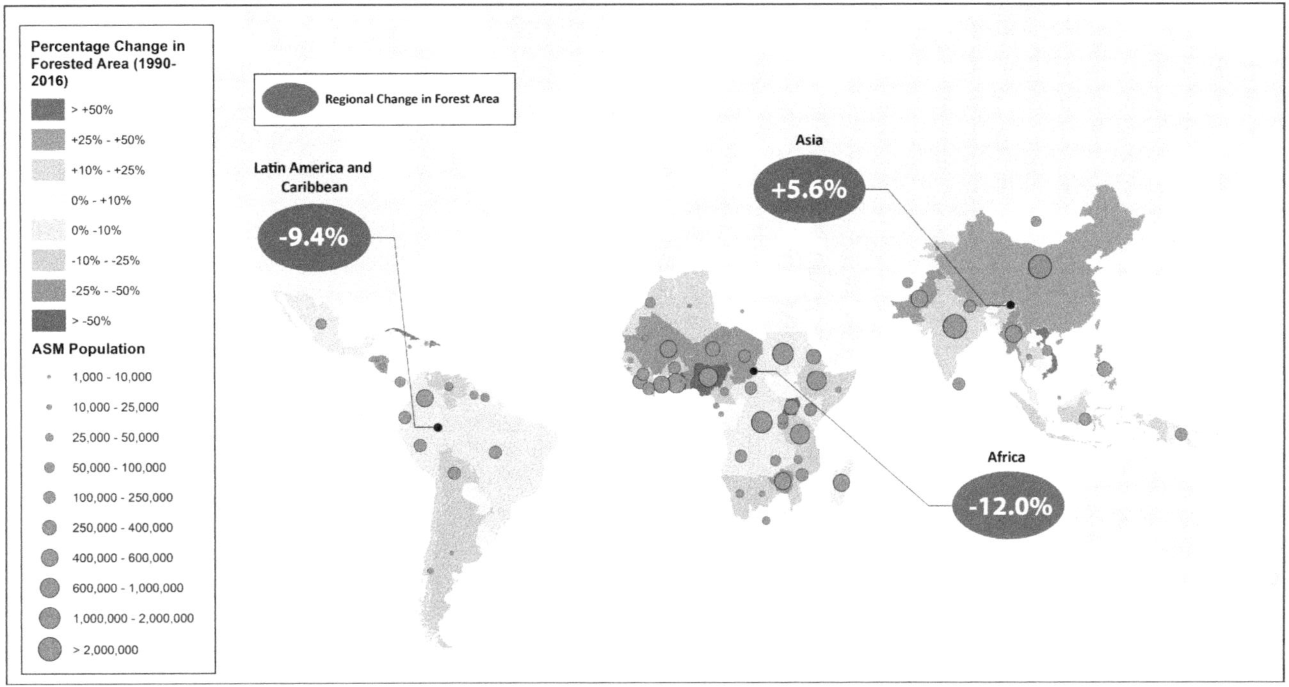


**FIGURE 15.4** Global rates of forest area as a proportion of total land area (km$^2$) between 1990–2016 compared against global ASM populations. Map depicts SDG 15 Indicator 15.1.1 trends in Latin America and the Caribbean (LAC), Africa, and Asia. Changes in forest area show greatest declines in Africa (–12.0%) followed by LAC (–9.4%), indicating cumulative land use pressures contributing to forest loss. Figure does not imply causation by ASM activity. Forest gains in Asia (+5.6%), however, overall trends in Asia do not reflect sub-regional declines in SE Asia (ECOSOC, 2020). Data: World Bank 2015 (forest area sq. km) and CASM (2017). (Original work created by Marco Brambilla and Kirsten Dales.)

Ecosystems Restoration (2021–2030), as well as other multilateral agreements on international waters and sound chemicals management (i.e., Minamata Convention).

While land restoration and rehabilitation are not new concepts per se, these have gained recent popularity and political currency, viewed as a linchpin to achieving all other SDGs (IRP, 2019). Combating land degradation and restoring degraded landscapes are indeed urgent global priorities to protect terrestrial ecosystems and the services that enable life on land. Approximately, 65% of Paris Agreement signatories have committed to restoring ecosystems (Seddon et al., 2019) and although there is scientific consensus on the critical importance of landscape restoration for climate mitigation efforts, there is growing concern over political blind-spots toward non-forested biomes despite significant carbon and biodiversity gains (Temperton et al., 2019; Seddon et al., 2019). Similarly, mining regeneration in tropical savannahs, fragile drylands, mountains, and rangelands are poorly documented in restoration science but of value to tackle pervasive global challenges of land degradation and desertification (Okpara et al., 2018).

As mining occurs in nearly all terrestrial biomes and important biodiversity areas on our planet, and ASM has limited capacity to avoid or reduce negative impacts, integrated landscape approaches *must* ensure no ecosystem is left behind in mine rehabilitation and closure planning. For example, ASM commonly occurs in tropical savannah and shrub-dominated lands across sub-Saharan Africa and elsewhere, but degradation can be difficult to attribute mining due to multiple land-uses, overgrazing, or patchy tree cover. However, tropical grasslands are resilient to land-use change as their recovery time and climate mitigation benefits can be realized on rapid time scales compared to tropical forest ecosystems (Goldstein et al., 2020). Consequently, tropical forests, forested wetlands (peatlands), mangroves, marshes, and other less charismatic grassy biomes must comprise the landscape restoration agenda to successfully achieve SDG target 15.3, 'Land Degradation Neutrality (LDN)'. Sustainable land management has become the focus of much of UNCCD's work, where the of quality of land resources to sustain ecosystem services and enhance food security are managed for stable equilibrium or increased within specified temporal and spatial scales and ecosystems (UNCCD, 2013; 2016). Thus, LDN offers a useful framework to address the broader impacts of mining, across a diverse range of ecosystems, habitat types and land-use scenarios.

Indeed, reversing land degradation from mining has important links to SDG 15.3 (LDN) as well as SDG 15 target 15.1 (terrestrial ecosystem restoration), 15.2 (forest restoration), 15.4 (mountains), 15.5 (biodiversity), 15.6 (genetic conservation), 15.8 (invasive species management), 15.9 (biodiversity values during mine life), and 15.a.b.c. (mobilizing finance for nature).

Unfortunately, legacies of mine abandonment are a harsh reality and land restoration and rehabilitation are not equal. In many cases, the most effective use of resources may not be to attempt to restore mining-impacted landscapes to pre-mining conditions, but to rehabilitate lands to stable, non-toxic landforms coupled with investments to reverse impacts on biodiversity corridors (Franks, 2015). Despite great need, there are persistent silos between ecological restoration, conservation, landscape ecology, and mining literature that undermine potential for the science-policy cohesion needed to facilitate mine rehabilitation and closure in ASM contexts.

Mine closure is a process of transitioning from active mining to an alternative land use, and involves environmental, social, and economic concerns. Reasons for mine abandonment are diverse and include (Dales and Ramasamy, 2019):

- Lack of mine closure policies or regulations, or requirements to remediate or rehabilitate abandoned mines;
- Ineffective enforcement of mine reclamation policies and/or minimal penalties for noncompliance;
- Lack of transparency in mine closure planning, funding mechanisms, or government oversight;
- Inadequate financial assurance, human resources or capacity to implement mine rehabilitation plans;
- Armed conflict, political instability or crisis, leading to an unscheduled halt of mining operations.

Environmental management guidelines and mine closure policies highlight the need for remediation, reclamation, rehabilitation, and restoration (R4) or a combination therein. These terms are often used interchangeably or vaguely defined, but differ in their definition, targets, biodiversity gains, and end goals (Table 15.2).

**TABLE 15.2**
**Remediation, Reclamation, Rehabilitation, and Restoration Definitions, Targets, Biodiversity Gains and End-Goals**

| Concept | Definition | Targets | Biodiversity Gains | End-Goal |
|---|---|---|---|---|
| **Remediation** | Remedy or correct a specific problem, often includes reducing pollution hazards at contaminated sites. | Soil decontamination | Low | Decontamination |
| **Reclamation** | Reclaim terrain stability and one aspect of ecosystem services, especially biogeochemical function of soils. | Soil stabilization and reduced erosion | Low–Medium | Physical terrain stability |
| **Rehabilitation** | Partial recovery of ecosystem services, dynamics, and biodiversity, optimizing post-mining land use goals. | Self-sustaining ecosystem | Medium–High | Post-mining land use |
| **Restoration** | Full recovery of ecosystem services, structures, dynamics, and biodiversity to pre-mining state. | Original ecosystem | High | Pre-mining land use |

*Source:* Dales and Ramasamy, 2019.

## CASE ANALYSIS

## DEFORESTATION, BIODIVERSITY LOSS, AND POLLUTION IN PERU'S MADRE DE DIOS

**Kirsten Dales**

**Paul Cordy**

Forests of the Peruvian Amazon are among the most biodiverse in the world. Tropical deforestation and degradation in Peru's Madre de Dios Department due to informal and illegal gold mining has drawn global attention, coupled with other historical processes of colonization and infrastructure investments. Like elsewhere in the Amazon, this process has also led to transgressions of the rights of indigenous peoples. Gold extraction from alluvial sediments has been occurring since the time of the Inca, within the region, but has caused ongoing land-use conflicts where mining, agriculture, and logging coexist.

The area is of global conservation importance and the epicentre of an ongoing alluvial gold rush spanning the Vilcabamba-Amboro Conservation Corridor, including portions of three Peruvian federally protected areas: the Tambopata National Reserve (2747 km$^2$), the Bahuaja-Sonene National Park (10,914 km$^2$), and the Amarakaeri Communal Reserve (4024 km$^2$). While these areas differ in their objectives in terms of conservation, perspectives are unified in their vilification of alluvial gold mining for devastating impact on eosystems, inland waters, and biodiversity, blaming mining for all land-use changes.

From 1985 to 2017, 95,751 ha of forest cover was lost due to gold mining activities in the Southeastern Peruvian Amazon. In the seven years from 2010–2017, 64,586 ha were lost to gold mining, more than double the total area lost in the preceding 26-year period (1985–2009, 31,165 ha), shown in Figure 15.5 (Espejo et al., 2018). The Interoceanic Highway connects the region, while navigable rivers and networks of secondary roads enable informal trade, human migration, and illegal tracking of mercury to process gold ores. Inherent challenges in avoiding, reducing, and reversing the impacts of recurrent gold mining rushes and their disruptions are further complicated by indirect drivers and root causes, including rural poverty, rising commodity prices, expanding road infrastructure, shifting political climate, and a near-continuous stream of illegal capital (Swenson et al., 2011; Kahhat et al., 2019).

Mining was the leading cause of deforestation in the region from 2010–2015 (60% vs. other land-uses). In the 2 years following however, other drivers of deforestation increased substantially, even though mining deforestation rates are at their highest levels ever, relative contribution of mining to 42 and 38% of total deforestation in 2016 and 2017, respectively (Espejo et al., 2018). Eviction of miners by force has been a typical response by local actors in attempts to avoid and reduce ongoing habitat destruction, illegal logging, and continued release of mercury-contaminated wastes into river systems. Police and military

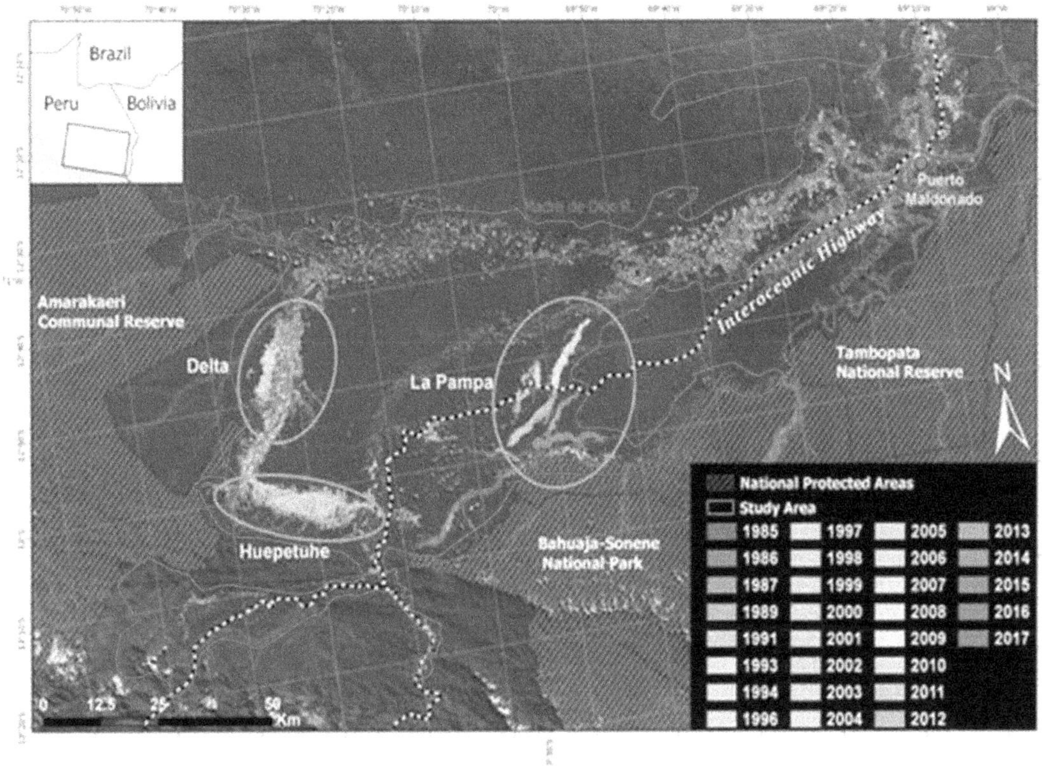


**FIGURE 15.5** Map of commodity-specific annual deforestation in the Southeastern Peruvian Amazon of Madre de Dios Department. Circles indicate deforestation due to artisanal and small-scale gold mining (ASGM) mining, with the year that land was first deforested displayed. A visual gradient of mining disturbance leading to increased deforestation impacts on deforestation, between two different time periods (1984–1985) and (2016–2017). Trends show earlier mining in the Huepetuhe and Delta, near the Amarakaeri Communal Reserve, with more recent mining intensity in La Pampa, bordering Bahuaja-Sonene National Park. Base map from ESRI. Open source image: Espejo et al., 2018.

forces have evicted over a thousand people and established a quarantine area that has effectively ended illegal logging and mining in La Pampa. However, this was achieved in part by pulling police forces from Puerto Maldonado to La Pampa's operations, with unintended consequences. For instance, the loss of illegal mining and timber harvest in areas with limited economic alternatives has led to a rise in crime in Puerto Maldonado, including robberies of eco-lodges and a negative impact on tourism. Some informal miners have instead started to operate illegally within legal concessions and indigenous lands, with severe impacts on the safety, security, and land tenure of the legal concession holders, inhabitants, and operators.

Local etiquette demands that miners (legitimate or not) pay royalties to indigenous communities. Since they cannot evict illegal miners for fear of punishment, disadvantaged communities accept the much-needed income. Few people outside these interior communities acknowledge the reality of these circumstances and accuse indigenous communities of participating in, benefiting from, and inviting illegal mining in their territories, further escalating tensions and eroding miner solidarity. All of these factors increase the costs of mining legally, whether through bribes that must be paid to local police to get fuel and tools into the field or through the fact that existing mineral tenure rules place responsibility for remediation on the concession holder of invaded lands. Needless to say, the local circumstance is complicated.

Mining was the leading cause of deforestation in the region from 2010–2015. In the past two years, however, other drivers of deforestation have increased substantially, even though mining deforestation rates are at their highest levels ever, reducing the relative contribution of mining to 42 and 38% of total deforestation in 2016 and 2017, respectively (Espejo et al., 2018). Shifting agriculture remains a significant driver of land-use change and deforestation, however illegal timber harvest is an ongoing threat as the forestry roads open new lands for settlers and agri-business. Annually, 12–28,000 ha of forest are destroyed illegally for mining. Still, the majority of this is done by miners in the process of formalization, who are trying to improve their practices and required to obtain forestry licenses for logging concessions, but rarely do (SERFOR (National forestry service) personal communication, 2019). Local authorities also require miners to develop, implement, and monitor a site remediation plan. Yet there is no legislation to enforce this or formal guidance on alluvial mine closure practices for miners to follow. SERFOR is currently working with MINEM (National Ministry of Mining and Energy) to provide the basic outline of mine closure for alluvial operations.

Undaunted, MINEM and DREM (Regional Energy and Mines Directorates) have formalized 105 miners since 2012 and aim to formalize another 150 (according to the MINEM special formalization task force in Puerto Maldonado) (El Comercio, 2019). According to local sources, the Governor of Madre de Dios aims to increase this number to 1100 in 2019–20. Yet most of the miners who have been formalized so far have taken no steps to comply with the responsibilities under the law that they have formally agreed to. Many lack access to finance, technical knowledge, or tools required to comply with existing regulations and many simply don't believe that there will be consequences. The latter attitude is exacerbated by the fact that the national regulatory bodies repeatedly extend amnesty periods and open up to new inscriptions of miners who want to formalize. Resident miners now live in fear of impending penalties that will accrue if a government enforcement agency decides to check up on them one day. It is well understood that miners must be given technical assistance in operations and closure. Yet outreach to date has mainly focused on mercury replacement with shaker tables, and other issues unrelated to environmentally sound chemicals and waste management, and local site remediation.

While individual mining operations in the region are small in scale and managed by individuals, they are interconnected by capital sources and can be viewed as working on behalf of larger enterprises, some of which finance dredging equipment. Unfortunately, there is simply no way to operate responsible mines using only the suction dredges that predominate in legal and illegal mining operations in Madre de Dios. Dredges extract alluvial sands at the bottom of mine pits, which expand as the edges are undermined, and the forest floor, including root bundles and soils, fold and slump into the pit. Mixing in organic materials not only devastates the ecological structure and function of fragile riverine habitats, but further exacerbates gold losses, which regularly exceed 60% – or more. Low recoveries ensure that miners recurrently scavenge previously mined lands when richer deposits are exhausted, or gold prices climb, thus destroying any natural regeneration or reforestation efforts and creating a disincentive for conservation purposes. Consequently, mining scale matters when considering the nature of impacts.

Indeed, much is said about the need for best practices and reforestation efforts in Madre de Dios. Efforts however are frustrated by the slow pace of formalization and lack of legitimate solutions or even basic guidelines for avoiding, reducing, or reversing impacts. The government of Peru and the Department of Madre de Dios have greatly increased their efforts to stop illegal mining and foster a responsible, formal, small-scale mining sector that can reduce poverty in a remote impoverished region. Nevertheless, there is a great deal that remains to be done, and despite laudable efforts by CINCIA (Centro de Innovacion Cientifica Amazonica, 2019) to develop reforestation techniques appropriate for mining degraded landscapes in the region, there is no budget for the daunting task of restoring wastelands that illegal mining has left behind.

## Key Points

- Narratives on sustainable development challenges and opportunities from those on the frontline in ASM communities are invaluable for understanding complex issues and shaping responses.

- In the densely forested tropical belts of the Southeastern Peruvian Amazon, deforestation has been attributed to illegal and informal gold mining, but is driven by underlying factors and root causes related to poverty, expansion of road infrastructure, shifting agriculture, and illegal timber harvest.
- In Peru's Madre de Dios Department, security of mineral tenure and allocation of land for mining are an ongoing challenge that undermine formalization.
- While individual mining operations in the region are small in scale and managed by individuals, they are interconnected by capital sources and should be treated differently to artisanal operators and individuals who generate comparatively little damage.
- Abandoned mines attract informal miners, where scavenging of waste and tailings discourages mine closure but can provide financial means for rehabilitation if positive incentives are in place.
- Unregulated ASM disrupts mine rehabilitation and closure planning, and where miners are not consulted, site invasion is inevitable.
- Reprocessing of tailings can (in theory) mobilize resources and potentially finance sustainable forest management, however, miners lack adequate incentives, training, and enforcement measures to advance such management, including reforestation and ecosystem rehabilitation.
- Peru's Madre de Dios shows that multifunctional mining landscapes are complex from ecology to economics and culture to politics, requiring multi-sectoral approaches for meaningful progress.
- Landscape approaches to optimize land allocation in rural economies have become *en vogue* in the last decade, as these approaches aim to balance competing land-uses in a way that is best for people and the environment. Their application to the ASM sector, however, has been largely absent to date.

## CONCLUSIONS

- Land and ecosystem degradation, driven by human activities, has undermined the well-being of at least 3.2 billion people worldwide (IPBES, 2018) and represents a serious threat to 'life on land'.
- Access to land with known mineral potential is critical for sustainable development of the ASM sector and multi-sectoral landscape approaches must aim to manage spaces, not people, within mining areas.
- Mining is a threat to terrestrial ecosystems, biodiversity, and major carbon sinks, including tropical forests, peatlands, mangroves, and grassy biomes (i.e., tropical savannas). While mining varies in production scale, degree of organization, and formality, ASM faces recurrent deficits in financial and human capital that interact to cyclically enhance poverty and resource degradation.

- Land restoration is a cost-effective way to mitigate the effects of climate change and biodiversity loss; however, SDG 15 targets and policy agendas indicate a 'blind spot' for grassy biomes, irrespective of non-forested ecosystems showing significant biodiversity and climate mitigation potential.
- The Land Degradation Neutrality (LDN) mitigation hierarchy to 'Avoid, Reduce, Reverse' equals the ambitions of SDG 15 to 'Protect, Sustainably Use, and Restore' within mining contexts.
- **Protect:** Despite positive global trends in expansion of protected areas (PAs) and financial assistance flowing toward species conservation and landscape restoration, terrestrial ecosystems, inland waters, and their biodiversity are being lost at rates that far exceed their conservation or sustainable use.
- **Manage:** Land restoration and rehabilitation can have significant co-benefits for all the SDGs, including positive and negative links that should be assessed for potential trade-offs and synergies.
- **Restore:** In mining, land restoration and rehabilitation are not equal. The most effective use of resources may not be to attempt to restore mining-impacted landscapes to pre-mining conditions but to rehabilitate lands to stable, non-polluting landforms coupled with investments to maintain connectivity of biodiversity corridors (see Franks, 2015).
- While the magnitude of impacts from industrial mines can be profound, the capacity of large-scale mining actors to mitigate and redress these far exceeds that of artisanal and small-scale miners.
- Mining can degrade ecosystems, fragment habitat, and threaten biodiversity. However, mining can finance alternative livelihoods that, over the long-term, may help to avoid biodiversity loss.*
- Abandoned mines attract informal miners, where scavenging of waste and tailings discourages mine closure but can provide financial means for rehabilitation (Dales and Ramasamy, 2019).
- Evidence suggests that informal or unregulated ASM can disrupt mine rehabilitation and closure, and often in cases where miners are not consulted in closure planning, site invasion is inevitable.
- Sustainable livelihood transitions for artisan and small-scale miners depend upon commodity prices, supply, demand, and incentives for conservation, restoration, and sustainable land-use.
- When properly engaged, educated, and empowered artisanal miners and mining communities can optimize local capacities toward sustainable livelihood and post-mining land-use transitions.
- The post-2020 biodiversity agenda, UN Decade on Ecosystem Restoration (2021–2030), UN Forum on Forests and LDN platform can jointly facilitate science-policy cohesion on mine closure in ASM zones and contexts.

---

* Recognizing there are shortfalls in all SDGs to adequately protect global biodiversity (see Zeng et al., 2020).

## REFERENCES

Abman, R. (2018). Rule of law and avoided deforestation from protected areas. *Ecological Economics* **146**(Dec 2017), 282–289. https://doi.org/10.1016/j.ecolecon.2017.11.004.

Afelt, A., Frutos, R., and Devaux, C. (2018). Bats, coronaviruses, and deforestation: Toward the emergence of novel infectious diseases? *Front Microbiol.* **9**, 1–5. https://doi.org/10.3389/fmicb.2018.00001.

Ali, S.H. *et al.* (2017). Mineral supply for sustainable development requires resource governance. *Nature* **543**(7645), 367–372.

Ali, S. (2018). The social ecology of artisanal mining: Between romanticisation and anathema. In: *Between the Plough and the Pick: Informal, Artisanal and Small-Scale Mining in the Contemporary World*, Lahiri-Dutt, ed., Australian National University Press, Canberra.

Alvarez-Berros, N.L., and Mitchell Aide, T. (2015). Global demand for gold is another threat for tropical forests. *Environmental Research Letters* **10**(1), 14006. https://doi.org/10.1088/1748-9326/10/1/014006.

Andrews, B.T., Elizalde, B., and Billon Le, P. (2017). *The Rise in Conflict Associated with Mining Operations: What Lies Beneath?*, 126. https://doi.org/10.13140/RG.2.2.36488.62720.

Baccini, A., Walker, W., Carvalho, L., Farina, M., Sulla-Menashe, D., and Houghton, R.A. (2017). Tropical forests are a net carbon source based on aboveground measurements of gain and loss. *Science* **358**, 230–234.

Bullock, J., Aronson, J., Newton, A., Pywell, R., and Benayas, J. (2011). Restoration of ecosystem services and biodiversity: Conflicts and opportunities. *Trends in ecology and evolution* **26**, 541–549. 10.1016/j.tree.2011.06.011.

Camizuli, E. *et al.* (2018). Trace metals from historical mining sites and past metallurgical activity remain bioavailable to wildlife today. *Scientific Reports* **8**(1), 3436.

Centro de Innovacion Cientifica Amazonica (2019). https://cincia.wfu.edu/en/publications/cincia-research-brief-no-4-reforestation-and-restoration-of-amazonian-landscapes-degraded-by-mining-species-analysis-and-solutions/.

Chomitz, K.M., Buys, P., Luca, G.D., Thomas, T.S., and Wertz-Kanounnikoff, S. (2007). At loggerheads? Agricultural expansion, poverty reduction and environment in the tropical forests. World Bank Policy Research Report.

Cowie, A., Orr, B., Castilla Sanchez, V., Chasek, P., Crossman, N., Erlewein, A., et al. (2018). Land in balance: The scientific conceptual framework for implementing land degradation neutrality. *Environmental Science and Policy* **79**, 25–35.

Curtis, P.G., Slay, C.M., Harris, N.L., Tyukavina, A., and Hansen, M.C. (2018). Classifying drivers of global forest loss. *Science* **361**(6407), 1108–1111. https://doi.org/10.1126/science.aau3445.

Dales, K. and Ramasamy, J. (2019). Mapping and Assessing Environmental Hazards of Abandoned Mines in Sub Saharan African Countries. UNESCO Publishing, Nairobi, Kenya. ISBN: 978-92-3-100360-8.

Davidson, J. (1993). The transformation and successful development of small-scale mining enterprises in developing countries. *Natural Resources Forum* **17**(4), 315–326.

de Castro Pena, J.C., Goulart, F., Wilson Fernandes, G., Hoffmann, D., Leite, F.S.F., Britto dos Santos, N., Soares-Filho, B., Sobral-Souza, T., Humberto Vancine, M., and Rodrigues, M. (2017). Impacts of mining activities on the potential geographic distribution of eastern Brazil mountaintop endemic species. *Perspectives in Ecology and Conservation.* https://doi.org/10.1016/j.pecon.2017.07.005.

Dominguez, S.R., O'Shea, T.J., Oko, L.M., and Homes, K.V. (2007). Detection of group 1 coronaviruses in bats in North America. *Emerging Infectious Disease* **13**: 1295–1300. https://doi.org/10.3201/eid1309.070491] PMID: 1825209845.

ECOSOC (2020). *Progress towards the Sustainable Development Goals Report of the Secretary-General.* ECOSOC, New York, USA.

Espejo, J.C., Messinger, M., Román-Dañobeytia, F., Ascorra, C., Fernandez, L.E., and Silman, M. (2018). Deforestation and forest degradation due to gold mining in the Peruvian Amazon: A 34-year perspective. *Remote Sensing* **10**(12), 1–17. https://doi.org/10.3390/rs10121903.

Fernandes, G.W., Goulart, F.F., Ranieri, B.D., Coelho, M.S., Dales, K., Boesche, N., ... Soares-Filho, B. (2016). Deep into the mud: Ecological and socio-economic impacts of the dam breach in Mariana, Brazil. *Natureza e Conservacao* **14**(2), 35–45.

Franks, D. (2015). *Mountain Movers.* Routledge, London, UK. https://doiorg.ezproxy.library.ubc.ca/10.4324/9781315884400.

Franks, D.M. (in press). Reclaiming the neglected minerals of development. *Extractive Industries and Society* (January), 1–8. https://doi.org/10.1016/j.exis.2020.02.002.

Getty, R., and Morrison-Saunders, A. (2020). Evaluating the effectiveness of integrating the environmental impact assessment and mine closure planning processes. *Environmental Impact Assessment Review* **82**(January), 106366. https://doi.org/10.1016/j.eiar.2020.106366.

Goldstein, A., Turner, W.R., Spawn, S.A., Anderson-Teixeira, K.J., Cook-Patton, S., Fargione, J., ... Hole, D.G. (2020). Protecting irrecoverable carbon in Earth's ecosystems. *Nature Climate Change* **10**(4), 287–295. https://doi.org/10.1038/s41558-020-0738-8.

Hattingh, R., Williams, D., and Corder, G. (2019). Applying a regional land use approach to mine closure: Opportunities for restoring and regenerating mine-disturbed regional landscapes. *Proceedings of the 13th International Conference on Mine Closure*, 951–968.

Hentschel, T., Hruschka, F., and Priester, M. (2003). *Artisanal and Small-Scale Mining, Challenges and Opportunities*, Projekt-Consult GmbH, MMSD project of IIED, 94 p.

Hilson, G. (2009). Small-scale mining, poverty and economic development in sub-Saharan Africa: An overview. *Resources Policy* **34**(1), 1–5.

Hilson, G., and Maponga, O. (2004). How has a shortage of census and geological information impeded the regularization of artisanal and small-scale mining? *Natural Resources Forum* **28**(1), 22–33.

Hosonuma, N., Herold, M., De Sy, V., De Fries, R.S., Brockhaus, M., Verchot, L., ... Romijn, E. (2012). An assessment of deforestation and forest degradation drivers in developing countries. *Environmental Research Letters* **7**(4). https://doi.org/10.1088/1748-9326/7/4/044009.

Huang, Y., Chen, Y., Castro-Izaguirre, N., Baruffol, M., Brezzi, M., Lang, A., Li, Y., Härdtle, W., ... Schmidt, G. (2018). Impacts of species richness on productivity in a large-scale subtropical forest experiment. *Science* **362**, 80–83.

IGF (Intergovernmental Forum on Mining, Minerals, Metals and Sustainable Development). (2017). *Global Trends in Artisanal and Small-Scale Mining (ASM): A Review of Key Numbers and Issues.* IISD, Winnipeg.

IPBES (2018). *Assessment Report on Land Degradation and Restoration.* United Nations Publishing. Accessible online at: https://ipbes.net/assessment-reports/ldr.

IPBES (2019). *Summary for Policymakers of the Global Assessment Report on Biodiversity and Ecosystem Services of the Intergovernmental Science-Policy Platform on Biodiversity and Ecosystem Services.* Díaz, S., Settele, J., Brondízio E.S., Ngo, H.T., Guèze, M., Agard, J., Arneth, A., Balvanera, P., ... Zayas, C.N. (eds.). IPBES Secretariat, Bonn, Germany.

IRP (2019). *Mineral Resource Governance in the 21st Century: Gearing Extractive Industries Towards Sustainable Development.* Ayuk, E.T., Pedro, A.M., Ekins, P., Gatune, J., Milligan, B., Oberle B., Christmann, P., Ali, S., Kumar, S. V, Bringezu, S., Acquatella, J., Bernaudat, L., Bodouroglou, C., Brooks, S., Burgii Bonanomi, E., Clement, J., Collins, N., Davis, K., Davy, A., Dawkins, K., Dom, A., Eslamishoar,

F., Franks, D., Hamor, T., Jensen, D., Lahiri-Dutt, K., Petersen, I., Sanders, A.R.D. *A Report by the International Resource Panel.* United Nations Environment Programme, Nairobi, Kenya.

IRP (2020). *Resource Efficiency and Climate Change: Material Efficiency Strategies for a Low-Carbon Future.* Hertwich, E., Lifset, R., Pauliuk, S., and Heeren, N. *A Report of the International Resource Panel.* United Nations Environment Programme, Nairobi, Kenya.

IUCN (2006). IUCN – CMP Unified Classification of Direct Threats. *IUCN* **2006**(June), 2–5.

Kahhat, R., Parodi, E., Larrea-Gallegos, G., Mesta, C., and Vázquez-Rowe, I. (2019). Environmental impacts of the life cycle of alluvial gold mining in the Peruvian Amazon rainforest. *Science of the Total Environment* **662**, 940–951. https://doi.org/10.1016/j.scitotenv.2019.01.246.

Mutemeri, N., Walker, J., Coulson, N., and Watson, I. (2016). Capacity building for self-regulation of the artisanal and small-scale mining (ASM) sector: A policy paradigm shift aligned with development outcomes and a pro-poor approach. *Extractives Industries and Society* **3**, 653–658.

Okpara, U.T., Stringer, L.C., Akhtar-Schuster, M., Metternicht, G.I., Dallimer, M., and Requier-Desjardins, M. (2018). A social-ecological systems approach is necessary to achieve land degradation neutrality. *Environmental Science and Policy* **89**(February), 59–66. doi:10.1016/j.envsci.2018.07.003.

Orr, B.J. et al. (2017). Scientific conceptual framework for land degradation neutrality. *A Report of the Science-Policy Interface.* United Nations Convention to Combat Desertification (UNCCD), Bonn, Germany. ISBN 978-92-95110-42-7.

Phillips, A. (2001). Mining and protected areas. *Mining, Minerals and Sustainable Development*, (October), 1–19.

Seal, R.R., II, Piatak, N.M., Kimball, B.E., and Hammarstrom, J.M. (2017). Critical mineral resources of the united states—economic and environmental geology and prospects for future supply: U.S. Geological Survey professional paper 1802. In: *Environmental Considerations Related to Mining of Nonfuel Minerals*, chap. B of Schulz, K.J., DeYoung, J.H., Jr., Seal, R.R., II, and Bradley, D.C., eds., U.S. Department of the Interior U.S. Geological Survey, p. B1–B16. ISSN 2330-7102 (online). https://pubs.usgs.gov/pp/1802/b/pp1802b.pdf.

Seddon, N., Turner, B., Berry, P., Chausson, A., and Giradin, C.A.J. (2019). Grounding nature-based climates solutions in sound biodiversity science. *Nature Climate Change* 9, 84–87.

Sonter, L.J., Herrera, D., Barrett, D.J., Galford, G.L., Moran, C.J., and Soares-Filho, B.S. (2017). Mining drives extensive deforestation in the Brazilian Amazon. *Nature Communications* **8**(1), 1–7.

Sonter, L.J., Ali, S.H., and Watson, J.E.M. (2018). Mining and biodiversity: Key issues and research needs in conservation science. https://doi.org/10.1098/rspb.2018.1926.

Swenson, J.J., Carter, C.E., Domec, J.C., and Delgado, C.I. (2011). Gold mining in the Peruvian amazon: Global prices, deforestation, and mercury imports. *PLoS ONE.* https://doi.org/10.1371/journal.pone.0018875.

Temperton, V.M., Buchmann, N., Buisson, E., Durigan, G., Kazmierczak, Ł., Perring, M.P., … Overbeck, G.E. (2019). Step back from the forest and step up to the Bonn Challenge: How a broad ecological perspective can promote successful landscape restoration. *Restoration Ecology* **27**(4), 705–719. https://doi.org/10.1111/rec.12989.

UNCCD (2013). *A Stronger UNCCD for a Land-Degradation Neutral World. UNCCD Secretariat Issue Brief.* UN Convention to Combat Desertification, Bonn, Germany.

UNCCD (2016). Land in Balance. The Scientific Conceptual Framework for Land Degradation Neutrality (LDN). Science-Policy Brief 02. September 2016, United Nations Convention to Combat Desertification (UNCCD), Science-Policy Interface. Bonn, Germany.

Valitutto, M.T., Aung, O., Tun, K.Y.N., Vodzak, M.E., Zimmerman, D., Yu, J.H., … Mazet, J. (2020). Detection of novel coronaviruses in bats in Myanmar. *PLoS ONE* **15**(4), 1–12. https://doi.org/10.1371/journal.pone.0230802.

Villegas, B.C., Weinberg, R., Levin, E., and Hund, K. (2012). *Artisanal and Small-Scale Mining in Protected Areas and Critical Ecosystems Programme (ASM-PACE):1–77.* Estelle Levin Limited and WWF, Cambridge, UK.

World Bank (2019). *Forest-Smart Mining: Identifying Good and Bad Practices and Policy Responses for Artisanal and Small-Scale Mining in Forest Landscapes*, 214.

Zeng, Y., Maxwell, S., Runting, R.K. *et al.* (2020). Environmental destruction not avoided with the Sustainable Development Goals. *Nat Sustain.* https://doi.org/10.1038/s41893-020-0555-0

GOAL 16

# Mining, Formalization and Peace, Justice and Strong Institutions

*Jorden de Haan*

## CONTENTS



> *Where resource exploitation has driven war, or served to impede peace, improving governance capacity to control natural resources is a critical element of peacebuilding.**

This chapter explores the relationship between mining and SDG 16 on peace, justice and strong and inclusive institutions. Given that the lion's share of security concerns associated with mining revolve around the – largely informal – ASM subsector, the chapter focuses specifically on ASM. Using ASM formalization as a framework of natural resource governance, it examines both its current and potential positive and negative contributions to the global realization of SDG 16 and its various dimensions. Moving beyond narratives of 'conflict minerals', this chapter argues that ASM formalization needs to be understood and prioritized as an integral part of peacebuilding, statebuilding and post-conflict reconstruction processes in conflict-affected and high-risk areas (CAHRAs), and beyond.

* Carolyn McAskie, Assistant Secretary-General for Peacebuilding Support, 2007, as cited in UN Environment, 2009.

## INTRODUCTION

The complex relationships between natural resources and peace, security, human rights and governance have been extensively documented in the last few decades, in both policy and academic circles (for example, see: Westing, 1986; Collier and Hoeffler, 1998; Le Billon, 2001; Alao, 2007; United States Institute of Peace, 2007; UN Environment, 2009). Indeed, according to UN Environment, between 1990 and 2009, at least 18 violent conflicts have been fuelled by the exploitation of natural resources, and over the last 60 years, at least 40% of all intrastate conflicts were linked to natural resources (UN Environment, 2009). With increased population growth and climate change, and resulting pressures on the competition over natural resources, conflicts over natural resources are likely to intensify in the future (EU Commission and High Representative, 2008).

But whereas historically, most analysis has focused on the *negative* impacts of natural resources on peace and security (e.g. natural resources' role in the outbreak of conflicts, financing and sustaining of conflicts, and impeding peace-making), the *positive* impacts of natural resources on peace and security have received little attention. Nevertheless, in recent years, natural resource management has been increasingly acknowledged as an opportunity for, and an integral part of, peacebuilding and post-conflict reconstruction processes (e.g. UN Environment, 2009; Jensen and Lonergan, 2012; Rustad, Lujala and Le Billon, 2012; Chatham House, 2013; Krampe, 2017).

Among the various resources that relate to peace and security, such as oil, gas, timber, livestock and water resources, *minerals* – and especially precious and strategic minerals that are 'lootable' (Westing, 1986) – arguably take centre stage. The role that 'blood diamonds' have played in fuelling the civil wars of Angola, Sierra Leone and Liberia is often used as a textbook example when discussing the natural resources–peace and security nexus. The role that other 'conflict minerals' (in particular gold, cassiterite, wolframite and coltan or '3TG') continue to play in the Eastern Democratic Republic of the Congo (DRC) and the African Great Lakes Region (GLR) has resulted in global media attention and development of international policy mechanisms. Specifically, it has given rise to the OECD Due Diligence Guidance for Responsible Supply Chains of Minerals from Conflict-Affected and High-Risk Areas* (hereafter: OECD Due Diligence Guidance) and the EU Conflict Minerals Regulation.† More recent examples include what Global Witness has described as Myanmar's 'Big State Secret': the jade trade in Kachin State, which is under complete control of military elites and drug lords (Global Witness, 2015); Colombia's small-scale gold sector, which is increasingly controlled by insurgent and criminal organisations involved in narcotics trade (Massé and Le Billon, 2017); and gold mines in Africa's Western Sahel, which have in the

* OECD, 2016. OECD Due Diligence Guidance for Responsible Supply Chains of Minerals from Conflict-Affected and High-Risk Areas: Third Edition.

† European Union, 2017. EU Conflict Minerals Regulation. Official Journal of the European Union, 60, 17 May 2017.

last few years been increasingly targeted by extremist groups and sparked recent media attention.*

Evidently, there are multiple links between mining and SDG 16 on peace, justice and strong institutions. In appreciation of mining's positive and potential impacts, the World Economic Forum and UNDP's *Mapping Mining to the Sustainable Development Goals: An Atlas* (hereafter: *Atlas*), describes these links as follows:

> Mining can contribute to peaceful societies and the rule of law by preventing and remedying company-community conflict, respecting human rights and the rights of indigenous peoples, avoiding illicit transfers of funds to public officials or other persons, ensuring transparent reporting of revenue flows, and supporting the representative decision-making of citizens and communities in extractives development. (UNDP, 2016, p. 2)

The *Atlas*, not unlike the available academic contributions that have to date investigated the links between mining and SDG 16, among other SDGs (de Mesquita, Xavier, Klein and Matos, 2017; Monteiro, da Silva and Neto, 2019), focuses exclusively on the large scale mining (LSM) sector, arguing that the artisanal and small-scale mining (ASM) sector warrants a separate guide in relation to the SDGs (UNDP, 2016, p.16). However, as Hilson and Maconachie (2019) have recently argued, this is a critical oversight, as it discourages policymakers from integrating ASM into national and regional development frameworks. Ali (2018) and the World Bank (2019) recently also underscored the need to link ASM to the SDGs (Ali, 2018; World Bank, 2019). This chapter seeks to address an important part of this gap by focusing exclusively on ASM in relation to SDG 16 – arguably the most challenging and most crosscutting of all SDGs (a forthcoming policy paper analyzes ASM in relation to all SDGs (de Haan, McQuilken and Dales, forthcoming)). In the author's perception, there are at least four reasons why this subsector merits an in-depth analysis here, which are related to four important elements of SDG 16: security, peacebuilding, human rights/justice and effective governance. These issues are briefly highlighted below, while an in-depth analysis of ASM formalization in relation to SDG 16 follows thereafter.

First, from a security perspective, although LSM is indeed associated with violent conflict, the lion's share of security concerns associated with the mining sector revolves around ASM. This point has recently been underscored by a quantitative study in the DRC, which demonstrates that armed groups mainly target ASM mines for financing their activities, while LSM companies are much less affected by them (Stoop, Verpoorten and van der Windt, 2019). A recent mapping of armed groups in the Eastern DRC by the International Peace Information Centre (IPIS) corroborates this point (IPIS, 2019). The study further shows that LSM companies' involvement in local conflicts (in the DRC) is broadly limited to conflicts with host communities and local artisanal miners (Stoop et al., 2019), a finding that is echoed in analyses in other parts of the world (UNDP, 2016; Andrews, Elizalde and Le Billon, 2017;

* See for example a recent Special Report by Reuters: Lewis, D. and McNeill, R. (2019, November 22). How jihadists struck gold in Africa's Sahel. Retrieved from www.reuters.com/investigates/special-report/gold-africa-islamists/

Mesquita, Xavier, Klein and Matos, 2017; Monteiro, da Silva and Neto, 2019). The strong association between ASM and insecurity could be explained by the obvious fact that the sector's high degree of informality renders its workforce vulnerable to external threats, as there is little formal oversight and workers have difficulties in claiming their rights and accessing grievance mechanisms. Verbrugge and Geenen (2019) further make the point that political crises, conflicts and the breakdown of political and economic structures have contributed to the expansion of ASM worldwide, arguing that ASM fares much better (than LSM and other sectors) in politically unstable contexts as the subsector is more in tune with local land tenure arrangements and is less affected by formal regulatory change (Verbrugge and Geenen, 2019). Indeed, the largely informal sector tends to flourish in fragile areas that have been affected by conflict, are considered as high-risk and/or are characterized by poor governance. This can be easily observed in the map in Figure 16.1, which overlays the 2019 State Fragility Index with ASM workforce estimates.[*,†]

While this map indicates larger ASM workforces in more fragile countries as a whole, there is, in addition to the cited IPIS maps of the DRC, also visual documentation for Colombia and Burkina Faso that illustrates the geographical overlap between ASM activity and the presence of armed groups *within* countries (for Colombia, see Massé and Le Billon, 2017: 4; for Burkina Faso, see Lewis and McNeil, 2019).

Second, when considering the socioeconomic dimension of peacebuilding and post-conflict reconstruction, ASM plays a vital role, even in its informal state. In short, based on conservative estimates, the ASM sector employs about five to six times more people than the LSM sector.[‡] The ASM workforce also includes a larger share of women (Hinton, Veiga and Beinhoff, 2003) and youth (Hilson and Osei, 2014; Maconachy, 2017) who play vital roles in peacebuilding processes (discussed later in this chapter). Moreover, while the sector provides *direct* livelihoods for more than an estimated 40 million people, it can be estimated to support another 120–280 million *indirect* livelihoods when considering the sector's positive spillover effects on rural economies.§

Third, as a result of their informality, ASM miners are particularly vulnerable to enduring human rights abuses and injustices. Cases of exploitation (see for example: de Haan and Geenen, 2016), extortion (e.g. Garrett, Sergiou and Vlassenroot, 2009), the (disputed) 'Worst Forms of Child Labour' (e.g. ILO, 2005) are widely reported in the ASM sector. Fourth, in terms of governance, ASM is plagued by incidents of bribery and corruption (see for example: Crawford and Botchwey, 2017; Hunter, 2019) and it has been argued that the prolonged informality of the sector in itself is, for a large part, a manifestation of poor governance (Hilson, 2013; Hilson and McQuilken, 2014; Hilson, Hilson, Maconachie, McQuilken and Goumandakoye,

* The full range of security, economic, political and social indicators of the Fragile States Index can be found here: https://fragilestatesindex.org/indicators/

† The ASM workforce estimates have been taken from the Artisanal and Small-scale Mining Knowledge Sharing Archive. http://artisanalmining.org/Inventory

‡ An estimated 40.5 million people were estimated to be directly engaged in ASM in 2017, up from 30 million in 2014, while only 7 million people were working in industrial mining in 2013 (IGF, 2017).

§ Using multiplier effects of three to seven, including both the lowest and highest range of multiplier effects that have been documented for the sector (World Bank, 2019).

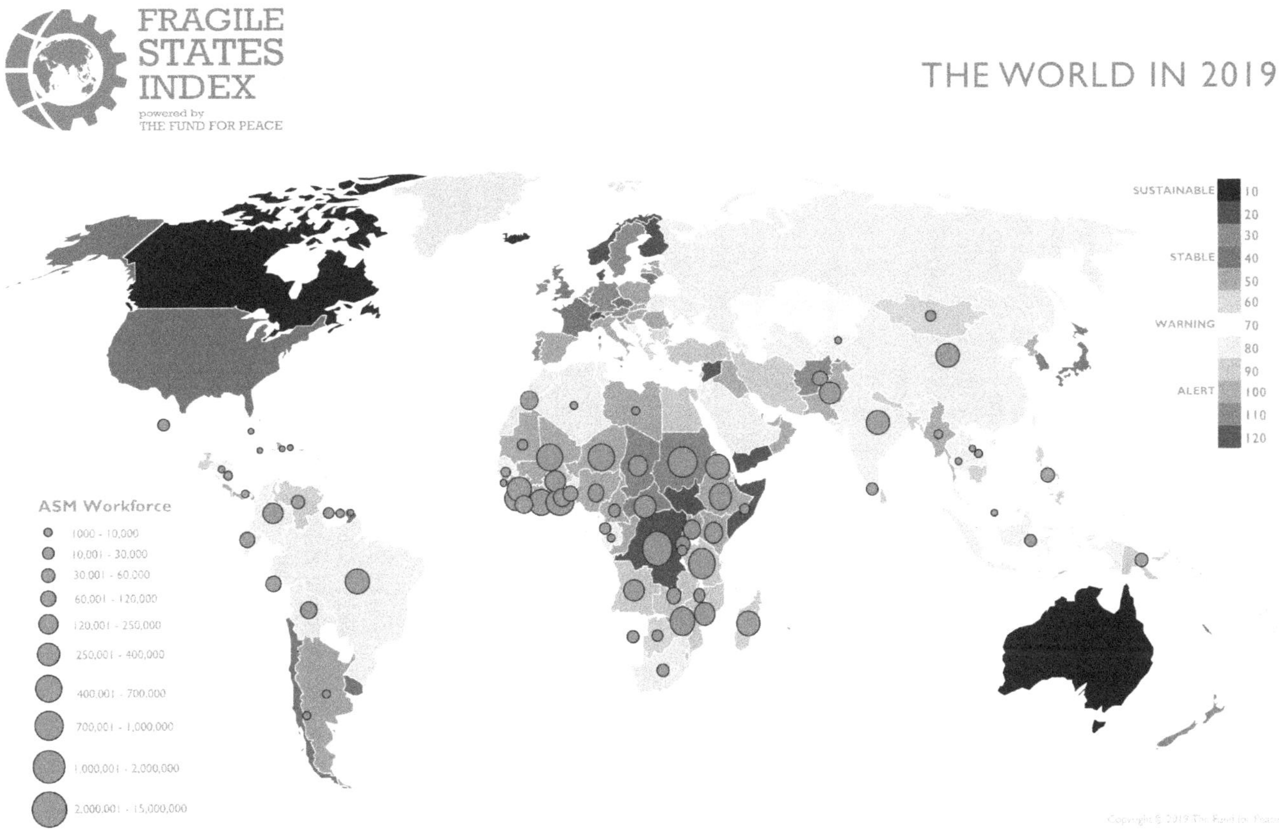


**FIGURE 16.1** 2019 Fragile States Index vs. estimated ASM workforce. Figure developed by Jorden de Haan and Imran Ahmed Mohamed, inspired by ASM Program Strategy, Canadian International Resources and Development Institute (CIRDI), developed by Kirsten Dales et al. Used with permission.

2017). Conversely, ASM formalization holds strong potential for redressing human rights abuses and advancing good governance.

Against this backdrop, this chapter proceeds by examining the interlinkages between ASM formalization and SDG 16. But before this can be done, both ASM formalization and SDG 16 need to be conceptualized and unpacked. Therefore, Section 1 reviews the key components of the formalization process, and introduces a human rights-based approach to ASM formalization. Subsequently, Section 2 unpacks SDG 16 and its identified key dimensions of peace, justice and inclusion. After that, Section 3, the core part of this chapter, analyses the interlinkages between ASM formalization and each of these three dimensions. This section draws upon some of the author's field work and professional work experiences in Sierra Leone and the DRC, but also includes references to the wider literature so as to maintain a global perspective. The chapter is concluded with a brief reflection on how policy-makers could best move forward in leveraging ASM formalization's full potential in enhancing peace, justice and inclusive institutions.

## SECTION 1. UNPACKING ASM FORMALIZATION

As has become clear from the preceding section, many of ASM's negative associations are 'expressions' of the sector's informality (Hilson, 2013). As such, and despite the disappointing progress that has been made in the last few decades (Hilson and McQuilken, 2014), there is a growing consensus in policy circles that formalization must be part of any strategy to develop the ASM sector (ILO, 1999; World Bank, 2009; UN Environment, 2012; UNITAR and UN Environment, 2018). Indeed, international (e.g. the OECD Due Diligence Guidance and the Minamata Convention on Mercury) and regional policy mechanisms alike (e.g. the International Conference on the Great Lakes Region's (ICGLR) Regional Initiative against the Illegal Exploitation of Natural Resources (RINR)) promote formalization as an essential step towards mitigating the sector's negative impacts and unlocking its full development potential. This mirrors a more general push for formalization of the informal sector, which is reflected in ILO's official Recommendation (204) concerning the Transition from the Informal to the Formal Economy,* and SDG 8 on Decent Work and Economic Growth, which equally encourages formalization. There is similar growing consensus in academic circles about the importance of formalization (Siegel and Veiga, 2009; Maconachie and Hilson, 2011; Hilson et al, 2017), although there are also reservations that top-down formalization efforts tend to benefit elites and can perpetuate or exacerbate inequalities (Geenen, 2012; Verbrugge, 2015; de Haan and Geenen, 2016), and have limited impacts if they are not coupled with education (Marshall and Veiga, 2017). It has thus been argued that formalization can only be a vehicle for development if it is approached in an *inclusive* and *comprehensive* manner (UNITAR and UN Environment, 2018; de Haan, 2019). Likewise, the author

* ILO, Recommendation 204 concerning the Transition from the Informal to the Formal Economy, adopted by the Conference at its 104th session, Geneva, 12 June 2015.

argues (once more, but this time in more depth*) that ASM formalization can be a vehicle for peace, justice and inclusion, but only if it is approached in such a manner.

With the conceptualization that follows, the author responds to Chapter 8 of this book, where Hilson (2021) highlighted the need to package and rebrand ASM formalization in more concrete development lingo which resonates more powerfully with decision-makers, as well as to Marshall and Veiga (2017), who have provoked the international community to 'get off the formalization train' – by proposing a framework for getting the train back on track towards peace, inclusive and sustainable development and justice.

While there are multiple interpretations of what ASM formalization entails exactly, there is consensus on the fact that formalization must be understood as a *process* (Siegel and Veiga, 2009) that includes multiple dimensions and goes *beyond legalization* (Verbrugge and Besmanos, 2016). Various IOs and INGOs have started further unpacking the ingredients of formalization. Analysis documents by the Alliance for Responsible Mining (ARM, 2011) and UN Environment (UN Environment, 2012) have provided particularly constructive inputs in this regard. Most recently, and perhaps most comprehensive to date, the UNITAR and UN Environment *Handbook for Developing National ASGM Formalization Strategies within National Action Plans* (hereafter: *Formalization Handbook*), which has been drafted by the author in collaboration with experts from IOs, governments, academia and INGOs, has provided a detailed conceptualization of the formalization process. It defines ASM formalization as follows:

> A process that ensures that AS[G]M actors are licensed and organized in representative entities that represent their needs; policies are implemented, monitored, and enforced; and AS[G]M actors receive technical, administrative, and financial support that empowers them to adhere to requirements prescribed by national regulations. (UNITAR and UN Environment, 2018, p. 17)

Essentially, formalization can be understood as a process of building both governments' and ASM actors' (understood as miners, traders and other supply-chain actors) capacity that progressively enables the latter to be in compliance with national regulations and international standards, and ultimately enables them to enter the global market.

Deriving from the above definition, the handbook distinguishes five *dimensions* of the formalization process: legal, institutional, socio-economic, geo-environmental and financial. Moreover, building on ideas from McQuilken and Hilson (2016), it introduces the *formality spectrum*, which helps to appreciate the many shades of grey that exist between 'formal' and 'informal', and illustrates increased degrees of formality as progress is made in the various components of formalization process. The *Handbook* then proceeds with discussing six *key components* of the

* This chapter builds on Section 3.4 of the UNITAR and UN Environment *Formalization Handbook*, which (briefly) discusses ASGM formalization in CAHRAs and introduces some of the points that are further elaborated here. For transparency, it must be noted that the author of this chapter is also the lead author of the *Formalization Handbook*.

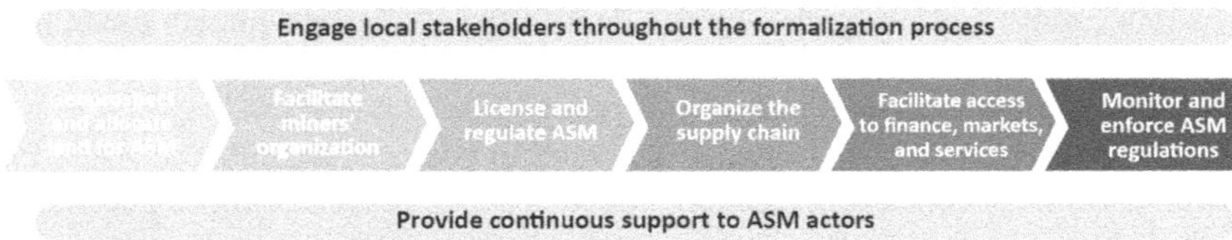


**FIGURE 16.2** Key components of the ASM formalization process.

formalization process, which are illustrated in Figure 16.2.* Under each component, both key steps and optional steps are introduced, providing flexibility for addressing country-specific and crosscutting issues (e.g. facilitating ASM-LSM co-existence, or managing ASM in protected areas).

While there is no space to discuss the concepts of the *Formalization Handbook* in more depth, a final element that merits a brief reflection in this chapter is the *Human Rights-Based Approach to ASM Formalization* (see Figure 16.3). Starting from the premise that ASM formalization can be a vehicle for development or for sustaining or exacerbating inequalities, depending on what approach is taken, the handbook introduces this human-centred approach to steer governments towards the latter.

Building on an approach introduced by the Sustainable Artisanal Mining (SAM) project funded by the Swiss Agency for Development Cooperation (SDC),† it focuses on the rights and duties of both ASM actors and the government. Moreover, as illustrated by the arrows in Figure 16.3, central to this approach is participation and consultation of ASM actors in a bottom-up rather than a top-down formalization process. Furthermore, as illustrated with the lower circle – and in line with observations that formalization needs to be grounded in local realities (Hison et al., 2017) and absorb existing customary practices (Siegel and Veiga, 2009) – it recognizes that the ASM sector is embedded in the local economy and local social order. Finally, as per the *UN Statement of Common Understanding on Human Rights-Based Approaches to Development Cooperation and Programming*, it focuses specifically on those stakeholders who are most marginalized, excluded or discriminated against (such as women, youth and the lowest-ranking miners).

## SECTION 2. UNPACKING SDG 16 AND ITS DIMENSIONS OF PEACE, JUSTICE AND INCLUSION

SDG 16 aims to "promote peaceful and inclusive societies for sustainable development, provide access to justice for all and build effective, accountable and inclusive institutions at all levels". As a lesson drawn from the Millennium Development Goals (MDGs), it was included in UN 2030 Agenda for Sustainable Development on the premise that there can be no inclusive and sustainable development without democratic governance, peace and security and the rule of law, including protection

* Figures 16.2 and 16.3 are slight adaptations of figures 4 and 5 of the *Formalization Handbook* (UNITAR & UN Environment, 2018), which focused on ASGM rather than ASM. Used with permission.

† See: http://sam.mn/sustainable-artisanal-mining-project/

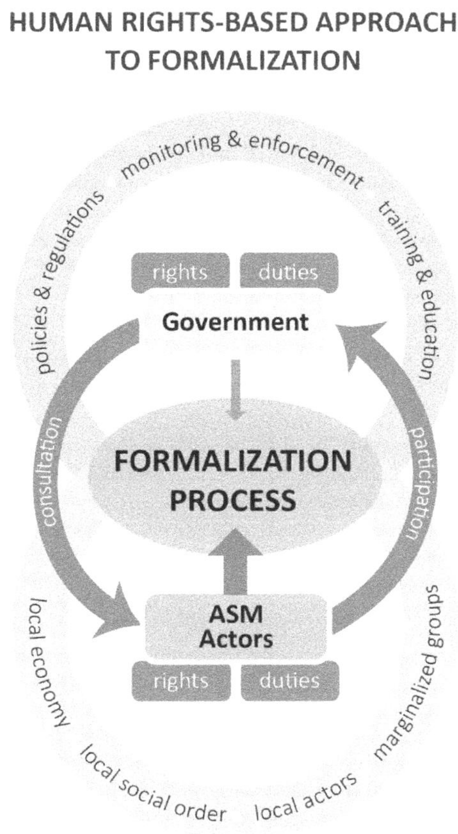


**FIGURE 16.3** A human rights-based approach to ASM formalization.

of human rights (The Global Alliance for Reporting Progress on Peaceful, Just and Inclusive Societies (hereafter: Global Alliance), 2019). Besides creating an enabling environment for the achievement of all other SDGs, the aims of SDG 16 are explicitly incorporated in Agenda 2030 as cross-cutting principles that should guide its implementation (e.g. 'leaving no one behind'), thus focusing not only on the *what* but also on the *how* by providing a human rights-based approach to implementing Agenda 2030. Obviously, the achievement of SDG 16 is a daunting task. It is not surprising that since the adoption of Agenda 2030 in 2016, each annual Report of the Secretary-General on the progress made on achieving the SDGs has consistently reported the uneven spread in progress towards achieving SDG 16, and underscored the critical need for renewed efforts towards this goal (United Nations Economic and Social Council, 2017, 2018, 2019, 2020).

The SDG in question includes many aspects, as illustrated by its 12 targets (see Table 16.1) and 23 corresponding indicators. However, as has also been argued by Whaites (2016), the targets' and indicators' causal relationships to the headline

**TABLE 16.1**
**SDG 16 Targets**

| | |
|---|---|
| **16.1** | Significantly reduce all forms of violence and related death rates everywhere. |
| **16.2** | End abuse, exploitation, trafficking and all forms of violence against and torture of children. |
| **16.3** | Promote the rule of law at the national and international levels and ensure equal access to justice for all. |
| **16.4** | By 2030, significantly reduce illicit financial and arms flows, strengthen the recovery and return of stolen assets and combat all forms of organized crime. |
| **16.5** | Substantially reduce corruption and bribery in all their forms. |
| **16.6** | Develop effective, accountable and transparent institutions at all levels. |
| **16.7** | Ensure responsive, inclusive, participatory and representative decision-making at all levels. |
| **16.8** | Broaden and strengthen the participation of developing countries in the institutions of global governance. |
| **16.9** | By 2030, provide legal identity for all, including birth registration. |
| **16.10** | Ensure public access to information and protect fundamental freedoms, in accordance with national legislation and international agreements. |
| **16.A** | Strengthen relevant national institutions, including through international cooperation, for building capacity at all levels, in particular in developing countries, to prevent violence and combat terrorism and crime. |
| **16.B** | Promote and enforce non-discriminatory laws and policies for sustainable development. |

goal cited earlier are questionable, and seem to fall short of grasping its full scope (Whaites, 2016). In recognition of this, the 'Pathfinders for Peaceful, Just and Inclusive Societies'* launched the Roadmap for Peaceful, Just and Inclusive Societies at the UN General Assembly in September 2017, which provides a shared vision for how 'SDG 16+' can be delivered. Based on Agenda 2030, the Pathfinders have identified three main domains of SDG 16 in this roadmap: peace, justice and inclusion. With the term 'SDG 16 plus', the Pathfinders convey that 24 targets under other SDGs also contribute to peace, justice and inclusion, and that they should equally be considered when discussing and pursuing SDG 16.† Telling examples include Target 5.5 (of SDG 5 – Gender Equality) on women's participation in leadership, or Target 4.7 (of SDG 4 – Quality Education) on a culture of peace and non-violence. While a detailed overview of the three domains of SDG 16 and their corresponding targets can be examined in the cited roadmap (Pathfinders for Peaceful, Just and Inclusive Societies, 2019, p. 16), Figure 16.4 provides a simplified overview of the three domains and a total of nine associated 'clusters' that include the various SDG targets (including all SDG 16 targets) that are related to each of the three domains.‡

* The Pathfinders for Peaceful, Just and Inclusive Societies are a group of member states, international organizations, global partnerships, and other partners committed to accelerating the delivery of the Agenda 2030 targets for peace, justice and inclusion (SDG 16+), hosted by the NYU Center on International Cooperation. https://www.sdg16hub.org/pathfinders

† Other prominent initiatives such as the Global Alliance have adopted this conception of SDG 16+.

‡ These nine 'clusters' have been developed by the Global Alliance (Global Alliance, 2019).

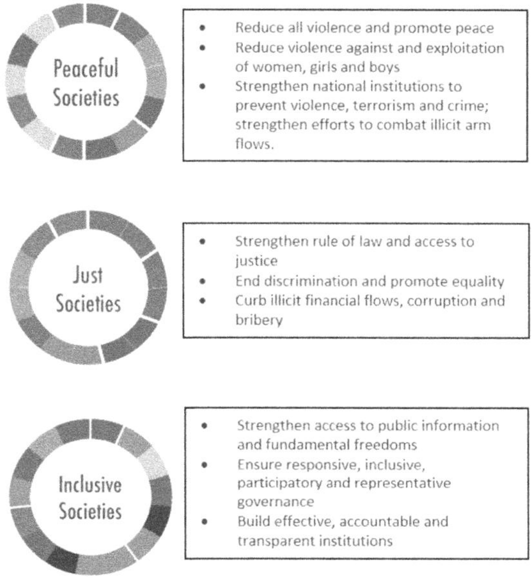


**FIGURE 16.4** Three domains of SDG 16+.

## SECTION 3. EXAMINING THE INTERLINKAGES BETWEEN ASM FORMALIZATION AND SDG 16

In appreciation of this more holistic conceptualization of SDG 16, the interlinkages between ASM formalization and SDG 16 will now be examined along the lines of the three mentioned domains, considering not only the SDG 16 targets, but also targets from other SDGs that have been identified to relate to these domains, where relevant. Whereas the *negative* associations between the informal ASM sector and peace, justice and inclusive governance have been well documented, this section will mainly explore the *positive* impacts of ASM formalization processes, in the spirit of the recent environmental peacebuilding literature. In doing so, the author recognizes that a large part of this potential is not yet realized because most governments do not (yet) address formalization from a human rights-based approach. Indeed, most governments take a rather top-down approach to the formalization process (Spiegel, 2015; de Haan and Geenen, 2016; Mutemeri et al, 2016; Verbrugge and Besmanos, 2016). At the same time, there are also indications that some countries are starting to adopt more bottom-up approaches. As discussed in Stylo, de Haan and Davis (2020), several countries that are developing NAPs have been conducting regular stakeholder engagement initiatives. For example, Sierra Leone has recently conducted workshops at both the national and the provincial level with a wide variety of stakeholders to map their needs and interests in ASGM (i.e. ASM gold), to develop a *national vision* for the sector and a comprehensive formalization strategy as per guidance of the *Formalization Handbook* (de Haan,

2019; Stylo, de Haan and Davis, 2020).* Such examples show that a shift in policy-making is thus possible. Still, it is in recognition of this caveat that this section now explores both the current and potential interlinkages between ASM formalization and SDG 16, including for situations where no human rights-based approach is adopted.

## ASM Formalization and Peaceful Societies

The domain of 'peaceful societies' includes three main objectives, as illustrated in Figure 16.4. Starting with the first objective, ASM formalization holds immense potential for reducing violence and promoting peace, and is already realizing a significant part of that potential. One direct way in which this occurs is through the establishment and implementation of supply-chain due-diligence systems (which can be understood as a part of component 4 of the formalization process: organization of the supply chain). For example, the iTSCi (the International Tin Supply Chain Initiative) mineral traceability and due diligence program for 3T minerals implemented by Pact in the DRC, Burundi, Rwanda and Uganda covers 2,200 mine sites across the four countries to date, positively impacting the lives of close to 80,000 miners by giving them access to the international market. If including the number of other people involved in the supply chain (transporters, middlemen, people who trade both minerals and other goods and services) along with immediate dependents of miners (estimated at 5 per miner), the total number of people benefitting from ITSCI is estimated to approach 1 million. In full alignment with the OECD Due Diligence Guidance, ITSCI records, monitors and reports risks along the supply chain, and facilitates their mitigation and resolution, aiming to create responsible and conflict-free 3T mineral supply chains that avoid contributing to armed conflict, human rights abuses, child labour or corruption.† Notwithstanding ongoing challenges in implementation and the risk of 'crime displacement effects',‡ the program has been effective in creating 3T mining areas that can be considered today as 'security hubs' in the region, by eliminating armed groups' control of ITSCI 3T sites (United Nations, 2018) and/or by preventing the capture of validated sites by armed groups (IPIS, 2019), thereby contributing to peace and security in insecure parts of the region (Pact, 2015).

More indirectly, but importantly, ASM formalization helps to address the root causes of the outbreak of violence. For example, in post-conflict Sierra Leone, where frustrations over unemployment, exclusion and a general loss of hope for the future among youth have been identified as some of the root causes of the civil war, ASM has provided ex-combatants with an alternative way of living (Maconachie and Hilson, 2011; Maconachie, 2017). At the same time, as the author has personally documented during fieldwork in Sierra Leone's Kono region (which hosts many ex-combatants), ASM miners have indicated that if they keep getting exploited by

---

* This national vision and ASGM formalization strategy are both included in Sierra Leone's NAP (EPA-SL, forthcoming).

† This can be verified in the latest quarterly ITSCI reports: www.itsci.org/status-report-public/

‡ i.e. displacing armed groups and the insecurity they bring to other natural resources in the Eastern DRC (Bafilemba, Mueller andand Lezhnev, 2014).

traditional authorities and do not receive more substantial support from the government soon to formalize their activities, they may 'return to the bush and continue to do what we did during the civil war …'.* Similarly, in Niger, ASM has offered people alternatives to unemployment, banditry or rebellion against the government (Grégoire and Gagnol, 2017). As such, and in light of the UN Secretary General's 'Plan of Action to Prevent Violent Extremism',† in which the provision of meaningful future prospects for youth is identified as a crucial step in the prevention of violent extremism, the comprehensive formalization of such miners' livelihoods can serve to address such frustrations by providing access to legal protection against exploitation, and provide more meaningful prospects for professionalizing their livelihoods in the future.

In a broader sense, ASM helps to address (or avoid) various forms of 'structural violence' (e.g. unequal distribution of benefits, exclusion) that are often the root causes of violent conflict, by contributing to the socio-economic dimension of peacebuilding processes.‡ As has been already illustrated in the introduction, ASM contributes significantly to direct and indirect livelihoods and has significant spillover effects on *local economies*. Moreover, a large part of the revenues benefits the *local population*.§ Furthermore, notwithstanding cases of informal taxation and exploitation, ASM also strengthens solidarity and promotes equality as miners from varying cultural backgrounds are easily accepted in ASM communities, irrespective of race and socio-economic status (Bryceson and Fisher, 2014; Bryceson and Geenen, 2016; Nkuba et al, 2019). As such, ASM's work culture contributes to fostering a culture of peace, which can be further leveraged by formalization. For example, by strengthening ASM cooperatives (Formalization component 2: facilitating miners' organization) on 'cooperative ideals' such as democratic governance, solidarity, equality and social justice, such entities can strengthen social cohesion.¶ Indeed, as reflected in the words of ILO's former Director-General, 'cooperatives can [also] be highly resilient in crisis and conflict and particularly valuable in building cooperation and solidarity and restoring dignity through self-help initiatives in situations of tension and post conflict situations'.** However, notwithstanding this *potential*, it must be noted that most ASM cooperatives are still a far cry from realizing such ideals in *practice* (see for example Fisher, 2007; de Haan and Geenen, 2016).

Moving to the second element of peaceful societies, ASM formalization can help to reduce violence against and exploitation of women, girls and boys. Women, who tend to be adversely affected by violent conflict, have been estimated to constitute about 30% of the global ASM workforce (Hinton, Veiga and Beinhoff, 2003) and have been argued to constitute more than that (Eftimie et al., 2012). Although most

* This quote is taken from a field study that the author undertook with EPA-SL in Kono and other regions of the country in January and February 2018. For an appreciation of the full context, readers can refer to EP-SL, 2018.

† United Nations General Assembly, A/70/674. Plan of Action to Prevent Violent Extremism.

‡ For a conceptualization of structural violence and different dimensions of peace(building), see Galtung, 1969.

§ For example, as has been documented in East Africa, at least 50% of the international value of gold is retained in the ASM communities where it is extracted (Barreto et al., 2018).

¶ See the International Cooperative Alliance's website for more on this: www.ica.coop/en

** http://www.ilo.org/public/english/bureau/dgo/speeches/somavia/2006/coopday06.pdf

women voluntarily pursue ASM as a livelihood because it is the best option available to them, they face a plethora of challenges – such as accessing land, mining groups, finance, technology and (formal) markets – that affects them disproportionally as a result of local gender dynamics (Hinton et al., 2003; Buss et al., 2017). Moreover, women operating in the informal ASM sector have been documented to fall victim to various forms of GBV, especially in conflict-affected areas (International Trade Union Conference, 2011; Rustad, Østby and Nordås, 2016). It has been argued that rather than seeking to remove women from this livelihood, efforts to protect such women from GBV should look to include them in mining reform initiatives (i.e. formalization processes) and empower them to improve their situation (Bashwira et al., 2014). This line of argument resonates with UN Security Council Resolution 1325 (2000) and the Women, Peace, and Security (WPS) agenda, which reaffirms the important role of women in conflict prevention, peacebuilding and post-conflict reconstruction. Indeed, as has been argued in the *Formalization Handbook*, the formalization of female ASM miners not only accommodates the regulatory oversight and access to protection that is required for reducing violence against women, but can also help to *empower* women in their capacity as peacebuilders.

The same can be said for youth, which dominates the ASM sector in Africa (Hilson and Osei, 2014) and likely worldwide. As can be deduced from the above example of young ex-combatants engaging in Sierra Leone's ASM sector, ASM formalization can go a long way in reducing miners' vulnerability to exploitation, and provide a meaningful alternative to picking up arms. Likewise, while UN Security Council Resolution 2250 on Youth, Peace, and Security* recognizes youth's important role in peacebuilding and post-conflict reconstruction, the formalization of young ASM miners equally presents an opportunity for empowering them as peacebuilders.

Beyond women and youth, a key concern that has driven the ASM formalization agenda is child labour. The ILO has long since documented children's involvement in ASM and the hazards they are exposed to (see for example ILO, 2005), and considers it one of the 'Worst Forms of Child Labour'.† As part of its approach to addressing this issue, ILO promotes the formalization of the ASM sector, as it serves to create an enabling environment for monitoring children's presence in the mines, and accommodates access to education and alternative sources of household income.‡ While it must be cautioned that child labour is a complex issue that requires addressing of its root causes (O'Driscoll, 2017), ASM formalization can thus help to create an enabling environment for reducing child labour and children's vulnerabilities to exploitation.

Finally, ASM formalization can help to strengthen national institutions to prevent violence, terrorism and crime, and strengthen efforts to combat illicit arms flows. At present, informal ASM and especially ASGM is associated with all of these issues. In Colombia, insurgent and criminal organisations involved in the narcotics trade wield control over informal small-scale gold mines to diversify and complement their

* United Nations Security Council, S/RES/2250 (2015).

† See ILO Convention 182 on the Worst Forms of Child Labour.

‡ ILO's approach to addressing child labour in ASM is further discussed in Section 3.2 of the cited *Formalization Handbook* (UNITAR and UN Environment, 2018).

sources of income, (re)investing gold proceeds in the coca business and vice versa to launder drug money (Massé and Le Billon, 2017). Likewise, in various African countries, criminal consortia have managed to assert control over the ASGM sector, not only to finance their activities but also for purposes of money laundering and tax evasion (Hunter, 2019). In Africa's Western Sahel, ASGM mines are increasingly targeted by terrorist organizations, such as the home-grown Ansarul Islam terrorist group which has targeted ASGM sites in Burkina Faso, and the Islamic State which has claimed several attacks in mining areas in Niger (OECD, 2018; International Crisis Group, 2019). In the DRC, artisanal diamond revenues have been documented to be used for financing arms trade (United Nations, 2002). Formalization can counter these issues in various ways. As was already discussed above, due diligence processes and formalization of ASM supply chains can constrain criminal, terrorist or insurgent groups in benefitting from ASM production and trade. Moreover, as is explained in more detail in the *Formalization Handbook* (Section 2.6), the component of monitoring and enforcing compliance with national regulations and international standards can help to identify and address the use of mineral proceeds for criminal purposes such as money laundering. Financial Intelligence Units (FIUs) can play an important role in this regard, as they receive, analyse and disseminate information relevant to money laundering, associated predicate offences and financing of terrorism; and can mobilize national security authorities.*

## ASM Formalization and Just Societies

The domain of 'just societies' includes three main objectives: i) strengthen the rule of law and access to justice; ii) end discrimination and promote equality; and iii) curb illicit financial flows, corruption and bribery. Starting with the first, the Global Alliance cautions that people who lack a legal identity are generally excluded from society and face difficulties in accessing public services, exercising the right to participate in civic processes and seeking justice for grievances (Global Alliance, 2019). A similar situation exists for informal ASM miners, who often have their rights violated by LSM companies, government officials and even armed groups, without possessing access to formal means of justice. Formalization can serve as such a means, by providing miners with mining titles that serve as a form of legal identity. As Siegel and Veiga already put it a decade ago, in their reflections on De Soto's concept of the dual economy and its meanings for ASM formalization: 'formal property rights are the basis of a miner's access to legal redress when rights are violated by a government or company, as they frequently are' (Siegel and Veiga, 2009, p. 52). As we have seen in Section 1, a human rights-based approach to formalization provides informal ASM actors with both rights and duties. These rights include both land tenure and labour rights, which can enable them to enjoy and exercise human rights as recognized in the Universal Declaration of Human Rights; for example, the right to recognition everywhere as a person before the law, or the right to an effective remedy for acts violating fundamental rights (UN General Assembly, Resolution 1948).

* Following Recommendation 29 of the Financial Action Task Force, many countries have established FIUs. For more information, see: https://egmontgroup.org/en/content/financial-intelligence-units-fius

Equally important, the duties may include, as per national regulations, things such as the (gradual) adoption of better mining practices or investments in local development. When governments enact their own duty of enabling ASM actors in carrying out such duties, the rule of law can be upheld.

Inclusive formalization processes further contribute to reducing discrimination and promoting equality. By legitimizing the livelihood of a historically marginalized workforce – including vast numbers of women (including widows and single mothers), poorly educated youth, orphans, former combatants, displaced people, refugees, economic migrants and, in general, some of the world's poorest people – and by facilitating its integration into the mainstream economy and society – ASM formalization can empower many of those who have been left furthest behind. In this regard, it is a promising trend that several countries have recently recognized ASM as a mining subsector in their regulatory frameworks for the first time,* transforming it from an extra-legal (often described locally as 'illegal') to an informal or legal activity; and that countries like Zimbabwe have recently 'decriminalized' informal ASM as part of their formalization efforts (Chipangura, 2019). The legal aspects (i.e. component 3) of the formalization process provide many other opportunities for developing policies and regulations for greater equality. For example, Sierra Leone has included in its National Action Plan's (NAP) formalization strategy steps to include gender-sensitive provisions in the amendment of its 2009 Mines and Minerals Act, as well as to streamline ASM in national gender policies and regulations (UN Environment, 2019; Environment Protection Agency Sierra Leone (EPA-SL), forthcoming).

Finally, ASM formalization can help to curb illicit financial flows (IFFs), corruption and bribery. IFFs, which can be understood as 'money illegally earned, transferred or used', have been identified in the vast majority of ASGM financial flows in Africa and Latin America (Global Initiative against Transnational Organized Crime and Levin Sources, 2017: 3–4). These flows tend to have a cyclical nature: while ASM gold may be smuggled over the border without paying taxes and could be used for purposes of money laundering, the proceeds from such activities may be reinvested in ASGM activity as a part of pre-financing arrangements between miners and gold buyers (idem). While in a context of widespread informality, not all types of IFFs should be outright presumed to be illegitimate (consider an example from Sierra Leone†), it is difficult to distinguish the legitimate from the illegitimate ones. However, what is clear is that formalization processes, albeit challenged by IFFs themselves, can help to manage IFFs and mitigate their impacts – for example, by making financial flows and the stakeholders involved in them

* For example, Uganda has just integrated ASM in its new Mining and Minerals Policy 2018; Mauritania has recently issued a Ministerial decree (Decree of 22 April 2016) to enable the licensing of ASM miners.

† Many of Sierra Leone's small gold traders smuggle gold over the Guinean and Liberian borders simply to avoid exorbitant export taxes. Once over the border, they receive US Dollars in exchange for their gold, which they may use to buy clothes and other goods which they sell back in Sierra Leone, or which they may reinvest in Sierra Leone's ASGM mines. Although this is illegal and can be considered as money laundering, the traders in question do not necessarily intend to engage in criminal activity or do harm in other ways; it just makes 'business sense' from their perspective. (This example is based on the author's field work in Sierra Leone, conducted with EPA-SL in January–February 2018.)

visible, by removing barriers for providing ASM miners with access to formal means of finance (replacing IFFs) and by creating incentives to engage in formal trade. Besides IFFs, the informal ASM sector is plagued by bribery and corruption, from Africa (Crawford and Botchwey, 2017; Hunter, 2019) to Latin America (Global Initiative against Transnational Organized Crime, 2016) and Asia (Global Witness, 2015). ASM actors without licences are easy targets for predatory (and often low-paid) government officials and traditional landowners. This leads to an often-cited reason why many informal ASM miners want to formalize: to be able to claim their rights, and to consolidate their bargaining position to deal with issues of bribery and exploitation.* Formalization can help protect ASM actors as they can demonstrate their possession of licences and payment of formal fees and can more easily seek legal redress; and it can further help to strengthen government institutions with whistleblowing and grievance mechanisms that can help to minimize bribery and corruption.

## ASM Formalization and Inclusive Societies

The domain of 'inclusive societies' includes three key objectives: i) strengthen access to public information and fundamental freedoms; ii) ensure responsive, inclusive, participatory and representative governance; and iii) build effective, accountable and transparent institutions.

The lack of publicly available information about the mining sector, such as census data, geological data and the allocation of mining titles, poses a key barrier to formalization (Hilson and Maponga, 2004; World Bank, 2019). Conversely, the process of gathering information and making it publicly available is an important part of the formalization process. For example, Sierra Leone's National Minerals Agency has established an Online Repository where people can log in to find data on all issued mineral rights, exploration, mining, dealers and exporters licences, and related payments.† Likewise, in its draft NAP, the DRC plans to make geological data, information about the availability of suitable land, and the rationale for the allocation of ASM and LSM areas publicly available through the Mining Cadre's (CAMI) website (Agence Congolaise de l'Environnement (ACE), forthcoming). While such repositories or mining portals do not always contain accurate and up-to-date information, and most *informal* ASM miners do not have the means to access them, they are a step in the right direction for making important information about the mining sector publicly accessible. Initiatives such as the World Bank–funded and Pact-implemented DELVE platform are leveraging and encouraging this trend.‡

If implemented from a human rights-based approach, ASM formalization can further contribute to the realization of fundamental freedoms and responsive, inclusive, participatory and representative governance. As explained in Section 1, central

---

* Based on the author's field work in South Kivu, DRC (with Agence Congolais de l'Environnement, July 2017) and Bo, Sierra Leone (with EPA-SL, March 2017).

† See https://sierraleone.revenuedev.org/dashboard

‡ See https://delvedatabase.org/

to this approach is participation and consultation of ASM actors – including especially those who are typically excluded – in a bottom-up process. Such initiatives enhance people's freedom of expression and participation in decision-making. For example, by organizing workshops where miners, traders and community members can voice their concerns about the current governance of their ASM sector and suggest new policy measures – as was recently done in Sierra Leone (de Haan, 2019) and the Eastern DRC (ACE, forthcoming). Moreover, while top-down formalization approaches, and in particular some countries' obligation to organize exclusively into ASM cooperatives, undermine miners' freedom of association (Mirindi, 2015: 593), more bottom-up approaches can instead enhance miners' freedom of association and assembly. For example, under Mongolia's SAM project, which also adopted a human rights-based approach, various consultations and workshops were organized, which allowed miners to engage in dialogue and establish their entities (unregistered partnerships, partnerships or cooperatives) according to their own pace and capacity. Furthermore, the Mongolian government facilitated a Mongolian Artisanal Miners Assembly in 2013, in which ASM miners have elected leaders of community-based 'miners' NGOs that serve to defend miners' interests. Subsequently, a National Federation for Artisanal Small-Scale Miners was established, which supports these miners' NGOs and represents ASM miners' interests to the Mongolian government (UNITAR and UN Environment, 2018: 29). Sierra Leone and the DRC have planned similar steps in their NAP formalization strategies, which include measures to facilitate the inclusion of women and youth and the assumption of leadership positions in such entities (EPA-SL, forthcoming; ACE, forthcoming). Such examples further underscore how inclusive and participatory approaches to ASM formalization can facilitate the social, economic and political inclusion of previously excluded citizens, and advance their representation at local and national levels.

As a final cluster of analysis, ASM formalization can help the formation of effective, accountable and transparent institutions. In the *Formalization Handbook*, the author already noted that formalization can enhance state-building and state reconstruction processes, and wider aspects of good governance (UNITAR and UN Environment, 2018). State-building can be understood as 'the creation of a government that has a monopoly of legitimate power and that is capable of enforcing rules throughout the state's territory' (Fukuyama, 2005). As has already been discussed in the previous section on just societies, formalization can help to enhance monitoring and enforcement of national regulations and strengthen the rule of law. Moreover, as part of the formalization process, many countries are decentralizing the issuance of mining licences and related tasks to provincial and district governments and, albeit to a more limited extent, building their financial and technical capacity for regulating the ASM sector (ARM, 2011; UNITAR and UN Environment, 2018). This can help, in the terminology of SDG target 16.6, to 'develop effective [...] institutions at all levels'. In addition, through field studies, stakeholder consultations and other activities, ASM formalization can help to foster mutual understanding between the government and ASM communities, thereby restoring trust and state legitimacy in areas where trust in the government has often been eroded.

State reconstruction is a similar concept, which includes elements such as the re-establishment of political institutions, the promotion of political participation and human rights, the provision of social services, and economic recovery (Englebert and Tull, 2008). During Mali's coup d'état in 2012, ASM played a critical role in funding rural institutions providing public services. While the central government's revenues collapsed, ASM revenues earned by miners and informally taxed by local authorities helped to fund local governments, schools and medical providers, enabling such institutions to proceed with their social services in a time of national crisis (Teschner, 2014). ASM formalization can further leverage and institutionalize such dynamics. For example, by facilitating formal taxation and the acquisition of foreign exchange earnings, and by stimulating ASM's spillover effects on local economies, ASM formalization can increase government resources and facilitate economic recovery in post-conflict settings more generally.

Finally, ASM formalization can enhance transparency and accountability. As discussed above, it can enhance public access to information about the extractives sector and transparency in mineral supply chains. Moreover, it can enhance transparency in revenue sharing in the sector, as it enables progress under the Extractives Industry Transparency Initiative (EITI), ITSCI and other initiatives. By allowing such public scrutiny, and by giving ASM actors a voice, relevant government departments can further be held to account in their governance of the ASM sector.

## CONCLUSIONS

This chapter has examined the interlinkages between ASM formalization and SDG 16 on peace, justice and inclusive institutions. It has started from the observation that the (largely informal) ASM sector is strongly related to poor governance, insecurity and abuse of human rights, and that such issues have – in combination with other drivers such as poverty and unemployment – given rise to the informal ASM sector in many parts of the world (Hilson, 2013; Hilson and McQuilken, 2014; Verbrugge and Geenen, 2019). Conversely, poor performance outcomes on SDG 16 – e.g. insecurity, IFFs, bribery and corruption – undermine the formalization process. Moreover, if approached in a top-down manner, ASM formalization can negatively affect peace, justice and inclusion. For example, as experiences from the DRC have shown, the imposition of cooperatives as the only legal mode of miners' organization can facilitate exploitation of labour (de Haan and Geenen, 2016); and ASM bans can trigger former ASM miners to (re)engage in criminal and rebel activities (Stoop, Verpoorten and van der Windt, 2019).

Moving beyond these negative (and better-documented) interlinkages, this chapter has explored both ASM formalization's current and potential *positive* impacts on SDG 16. Using a comprehensive conceptualization of 'SDG 16+', it has been demonstrated that ASM formalization has *great potential* to contribute to *all* of the identified clusters of SDG targets under SDG 16's three dimensions of peace, justice and inclusion. To date, a relatively small yet considerable part of that potential is already being realized. For example, by facilitating the demilitarization of certain mineral supply chains; by facilitating processes of state-building and state reconstruction;

and by legitimizing the livelihood of a historically marginalized workforce and facilitating its integration into the mainstream economy – thereby empowering many of those who have been left furthest behind. As such, this chapter has confirmed the positive role that natural resource management can play in peacebuilding and post-conflict reconstruction (UN Environment, 2009; Jensen and Lonergan, 2012; Chatham House, 2013), in the context of ASM.

However, as has become evident, the unrealized potential is much bigger. Governments could realize more of it by shifting part of their focus from legalization to other components of the comprehensive formalization process, and by adopting more inclusive and bottom-up approaches. In this respect, the human rights-based approach to formalization that has been introduced in the *Formalization Handbook*, and discussed here, can serve as a framework for understanding and enhancing ASM's contributions to peace, justice and inclusion. When such an approach is adopted, it becomes less about the *outcomes* of formalization processes – which can take decades to complete, if ever – and more about the *process* of formalization itself, which, from this lens, is fundamentally a process of state-building, peacebuilding and development.

To be sure, the global efforts that are required for the achievement of the ambitious SDG 16 go far beyond ASM formalization, and must include broader strategies for promoting issues such as democratic governance, inclusive development, access to justice, and security sector reform. Nevertheless, as has been pointed out before (OECD, 2018; UNITAR and UN Environment, 2018), and as this chapter has attempted to demonstrate, ASM formalization needs to be understood and prioritized as an integral part of such strategies, and correspondingly be integrated into national, regional and global frameworks of peace and security.*

## ACKNOWLEDGEMENTS AND DISCLOSURE STATEMENT

The author would like to thank all staff members of Pact's Mines to Market (M2M) program for their inputs during a discussion on ASM formalization and the SDGs, which have been insightful for shaping certain parts of this chapter. In particular, he would like to thank Mr Mickaël Daudin, program officer with Pact M2M, for his inputs regarding the ITSCI program. The author would also like to thank Mr Imran Ahmed Mohamed, IT assistant with Pact Kenya, for his support in (re)designing Figures 16.1–16.3 of this chapter. It should further be noted that Jorden de Haan currently works as a program officer with Pact, an international development organization specialized in ASM. Prior to this, he worked as a consultant with UNITAR, where he has coordinated the NAP projects in Sierra Leone, the DRC and Eritrea, and where he has led the authoring of the *Formalization Handbook* discussed in this chapter.

* Concrete steps and recommendations for integrating ASM into national and regional development frameworks are provided in de Haan, McQuilken and Dales (forthcoming).

## REFERENCES

Agence Congolaise de l'Environnement, forthcoming. Plan d'Action National pour réduire et si possible, éliminer l'utilisation du mercure dans l'Extraction Minière Artisanale et a Petite Echelle de l'or (EMAPE) en République Démocratique du Congo. *Final Draft Pending National Endorsement.*

Alao, A., 2007. *Natural Resources and Conflict in Africa: The Tragedy of Endowment* (Vol. 29). University Rochester Press.

Ali, S. H., 2018. The social ecology of artisanal mining: Between romanticisation and anathema. *The Pick, 117*, 3.

Alliance for Responsible Mining, 2011. Analysis for stakeholders on formalization in the artisanal and small-scale gold mining sector based on experiences in Latin America, Africa and Asia.

Andrews, T., Elizalde, B., Le Billon, P., Hoon Oh, C., Reyes, D., and Thomson, I., 2017. *The Rise in Conflict Associated with Mining Operations: What Lies Beneath.* Washington, DC: CIRDI, pp. 1–127.

Bafilemba, F., Mueller, T., and Lezhnev, S., 2014. *The Impact of Dodd-Frank and Conflict Minerals Reforms on Eastern Congo's Conflict.* Washington, DC: Enough Project.

Barreto, M. L., Schein, P., Hinton, J., and Hruschka, F., 2018. *The Impact of Small-Scale Mining Operations on Economies and Livelihoods in Low- to Middle-Income Countries.* PACT Inc and Alliance for Responsible Mining.

Bashwira, M.-R., Cuvelier, J., Hilhorst, D., and van der Haar, G., 2014. Not only a man's world: Women's involvement in artisanal mining in eastern DRC. *Resources Policy, 40*(1), 109–116.

Bryceson, D., and Fisher, E., 2014. Artisanal mining's democratizing directions and deviations. In: Bryceson, D. F., Fisher, E., Jønsson, J. B., Mwaipopo, R. (Eds.), *Mining and Social Transformation in Africa: Mineralizing and Democratizing Trends in Artisanal Production.* Oxford: Routledge, pp. 181–209.

Bryceson, D. F., and Geenen, S., 2016. Artisanal frontier mining of gold in Africa: Labour transformation in Tanzania and the Democratic Republic of Congo. *African Affairs, 115*(459), 296–317.

Buss, D., Rutherford, B., Hinton, J., Stewart, J., Lebert, J., Cote, G. E., Sebina-Zziwa, A., Kibombo, R., and Kisekka, F., 2017. *Gender and Artisanal and Small-Scale Mining in Central and East Africa: Barriers and Benefits (Working Paper GWP-2017-2).* Ottawa: Institute for the Study of International Development (IDRC).

Chatham House, 2013. Encouraging peace-building through better environmental and natural resource management. *Chatham House*, December 2013. https://www.chathamhouse.org/publications/papers/view/196436.

Chipangura, N., 2019. Towards the decriminalisation of artisanal gold mining in Eastern Zimbabwe. *The Extractive Industries and Society, 6*(1), 154–161.

Collier, P., and Hoeffler, A., 1998. On economic causes of civil war. *Oxford Economic Papers, 50*(4), 563–573.

Crawford, G., and Botchwey, G., 2017. Conflict, collusion and corruption in small-scale gold mining: Chinese miners and the state in Ghana. *Commonwealth and Comparative Politics, 55*(4), 444–470.

de Haan, J. S., and Geenen, S., 2016. Mining cooperatives in Eastern DRC: The interplay between historical power relations and formal institutions. *The Extractive Industries and Society, 3*(3), 823–831.

de Haan, J. S., McQuilken, J., and Dales, K. M., forthcoming. Mapping artisanal and small-scale mining to the Sustainable Development Goals. Forthcoming policy paper with the University of Delaware.

de Haan, J. S., 2019. Participatory governance as the means and the end of development. Engaging artisanal gold miners and traders in developing Sierra Leone's National Action Plan under the Minamata Convention on Mercury [magazine article]. *UN Special*, May 15, May edition 2019, Geneva. https://www.unspecial.org/2019/05/participatory-governance-as-the-means-and-the-end-of-development/.

ECOSOC, 2018. Progress towards the sustainable development goals: Report of the secretary-general, UN Doc E/2018/64. New York: ECOSOC. https://unstats.un.org/sdgs/files/report/2018/secretary-general-sdg-report-2018--EN.pdf.

ECOSOC, 2019. Special edition: Progress towards the sustainable development goals: Report of the secretary-general, UN Doc E/2019/68. New York: ECOSOC. https://unstats.un.org/sdgs/files/report/2019/secretary-general-sdg-report-2019--EN.pdf.

ECOSOC, 2020. Progress towards the Sustainable Development Goals: Report of the secretary-general, UN Doc E/2020/xxx. New York: ECOSOC.

Eftimie, A., Heller, K., Strongman, J., Hinton, J., Lahiri-Dutt, K., and Mutemeri, N., 2012. *Gender Dimensions of Artisanal and Small-Scale Mining: A Rapid Assessment Toolkit*. World Bank. https://openknowledge.worldbank.org/handle/10986/2731.

Englebert, P., and Tull, D. M., 2008. Postconflict reconstruction in Africa: Flawed ideas about failed states. *International Security*, *32*(4), 106–139.

EPA-SL, 2018. *The ASGM Overview of Sierra Leone*. Freetown, Sierra Leone: Environment Protection Agency.

EPA-SL, forthcoming. National Action Plan for Reducing Mercury Use in the Artisanal and Small-scale Gold Mining (ASGM) Sector in Sierra Leone. Final draft pending national endorsement.

EU Commission and High Representative, 2008. *Climate Change and International Security: Paper to the European Council*. S113/8. Brussels: European Council.

Fisher, E., 2007. Occupying the margins: Labour integration and social inclusion in artisanal mining in Tanzania. *Development and Change*, *38*(4), 735–760.

Fukuyama, F., 2005. Building democracy After conflict: 'Stateness' first. *Journal of Democracy*, *16*(1), 84–88.

Galtung, J., 1969. Violence, peace, and peace research. *Journal of Peace Research*, *6*(3), 167–191.

Garrett, N., Sergiou, S., and Vlassenroot, K., 2009. Negotiated peace for extortion: The case of Walikale territory in eastern DR Congo. *Journal of Eastern African Studies*, *3*(1), 1–21.

Geenen, S., 2012. A dangerous bet. The challenges of formalizing artisanal mining in the Democratic Republic of Congo. *Resources Policy*, *37*(3), 322–330.

Global Initiative Against Transnational Organized Crime, 2016. Organized crime and illegally mined gold in Latin America. April 2016, Geneva. https://globalinitiative.net/organized-crime-and-illegally-mined-gold-in-latin-america/.

Global Alliance for Reporting Progress on Peaceful, Just and Inclusive, 2019. Societies, 'enabling the implementation of the 2030 agenda through SDG 16+: Anchoring peace, justice and inclusion'. New York: United Nations. https://www.SDG 16hub.org.

Global Initiative Against Transnational Organized Crime and Levin Sources, 2017. Follow the money: A handbook for identifying financial flows linked to Artisanal and Small-Scale Gold Mining. https://www.levinsources.com/publications/giff-mapping-iffs-in-asgm.

Global Witness, 2015. Jade: Myanmar's "big state secret". Global Witness, October. https://www.globalwitness.org/en/campaigns/oil-gas-and-mining/myanmarjade/.

Grégoire, E., and Gagnol, L., 2017. Ruées vers l'or au Sahara: L'orpaillage dans le désert du Ténéré et Le Massif de l'Aïr (Niger). *EchoGéo*. http://journals.openedition.org/echogeo/14933 DOI:10.4000/echogeo.14933.

Hilson, G., 2013. "Creating" rural informality: The case of artisanal gold mining in sub-Saharan Africa. *SAIS Review of International Affairs*, *33*(1), 51–64.

Hilson, G., 2021. Promote sustained, inclusive and sustainable economic growth, full and productive employment and decent work for all. In: Parras, C., Ali, S., Lewis, B. (Eds.), *Mining, Materials, and the Sustainable Development Goals (SDGs): 2030 and Beyond*. Routledge.

Hilson, G., Hilson, A., Maconachie, R., McQuilken, J., and Goumandakoye, H., 2017. Artisanal and small-scale mining (ASM) in sub-Saharan Africa: Re-conceptualizing formalization and 'illegal' activity. *Geoforum*, *83*, 80–90.

Hilson, G., and Maconachie, R., 2019. Artisanal and small-scale mining and the Sustainable Development Goals: Opportunities and new directions for sub-Saharan Africa. *Geoforum*, *111*, 125–141.

Hilson, G., and Maponga, O., 2004. How has a shortage of census and geological data inhibited the regularization of artisanal and small-scale mining? *Natural Resources Forum*, *28*(1), 22–33.

Hilson, G., and McQuilken, J., 2014. Four decades of support for artisanal and small-scale mining in sub-Saharan Africa: A critical review. *The Extractive Industries and Society*, *1*(1), 104–118.

Hilson, G., and Osei, L., 2014. Tackling youth unemployment in sub-Saharan Africa: Is there a role for artisanal and small-scale mining? *Futures*, *62*, 83–94.

Hinton, J., Veiga, M., and Beinhoff, C., 2003. Women and artisanal mining: Gender roles and the road ahead. In: Hilson, G., Pub, A. A., Balkema, A. (Eds.), *The Socio-Economic Impacts of Artisanal and Small-Scale Mining in Developing Countries*. Netherlands: Swets Publishers.

Hunter, M., 2019. Pulling at golden webs. Combating criminal consortia in the African artisanal and small-scale gold mining and trade sector. *Enact*, Issue 8, April 2019. https://enactafrica.org/research/research-papers/pulling-at-golden-webs-combating-criminal-consortia-in-the-african-artisanal-and-small-scale-gold-mining-and-trade-sector.

ILO, 1999. Social and labour issues in small-scale mines: Report for discussion at the tripartite meeting on social and labour issues in small-scale mines. http://www.ilo.org/global/publications/ilo-bookstore/order-online/books/WCMS_PUBL_9221114805_EN/lang--en/index.htm.

ILO, 2005. Minors out of mining - partnership for global action against child labour in small-scale mining. https://www.ilo.org/ipec/areas/Miningandquarrying/WCMS_163749/lang--en/index.htm.

Intergovernmental Forum on Mining, Minerals, Metals and Sustainable Development (IGF), 2017. *Global Trends in Artisanal and Small-Scale Mining (ASM): A Review of Key Numbers and Issues*. Winnipeg: IISD.

International Crisis Group, 2019. Getting a grip on Central Sahel's Gold Rush. *Africa Report N°282*, November 13.

International Trade Union Conference (ITUC), 2011. Violence against women in Eastern Democratic Republic of Congo: Whose responsibility? Whose complicity? Brussels: International Trade Union Conference.

IPIS, 2019. Mapping artisanal mining areas and mineral supply chains in eastern DR Congo. Impact of Armed Interference and Responsible Sourcing, Antwerp.

Jensen, D., and Lonergan, S., 2012. Natural resources and post-conflict assessment, remediation, restoration, and reconstruction: Lessons and emerging issues. *Assessing and Restoring Natural Resources in Post-Conflict Peacebuilding*, pp. 411–461.

Krampe, F., 2017. Toward sustainable peace: A new research agenda for post-conflict natural resource management. *Global Environmental Politics*, *17*(4), 1–8.

Le Billon, P., 2001. The political ecology of war: Natural resources and armed conflicts. *Political Geography*, *20*(5), 561–584.

Lewis, D., and Mcneill, R., 2019. How jihadists struck gold in Africa's Sahel. *Special Report by Reuters*, November 22. https://www.reuters.com/investigates/special-report/gold-africa-islamists/.

Maconachie, R., 2017. Navigating the intergenerational divide? Youth, artisanal diamond mining, and social transformation in Sierra Leone. *The Extractive Industries and Society*, *4*(4), 744–750.

Maconachie, R., and Hilson, G., 2011. Artisanal gold mining: A new frontier in post-conflict Sierra Leone? *The Journal of Development Studies*, *47*(4), 595–616.

Marshall, B. G., and Veiga, M. M., 2017. Formalization of artisanal miners: Stop the train, we need to get off! *The Extractive Industries and Society*, *4*(2), 300–303.

Massé, F., and Le Billon, P., 2017. Gold mining in Colombia, postwar crime and the peace agreement with the FARC. *Third World Thematics: A TWQ Journal*. DOI:10.1080/23802014.2017.1362322.

de Mesquita, R. F., Xavier, A., Klein, B., and Matos, F. R. N., 2017. Mining and the sustainable development goals: A systematic literature review. *Geo-Resources Environment and Engineering (GREE)*, *2*, 29–34.

McQuilken, J., and Hilson, G., 2016. *Artisanal and Smallscale Gold Mining in Ghana. Evidence to Inform an 'Action Dialogue'*. International Institute for Environment and Development: London.

Mirindi, P. L., 2015. La coopérative minière: Instrument de l'ingérence étatique dans la liberté d'association des exploitants miniers artisanaux en République Démocratique du Congo? *KAS African Law Study Library*, *1*(3), 563–598.

Monteiro, N. B. R., da Silva, E. A., and Neto, J. M. M., 2019. Sustainable development goals in mining. *Journal of Cleaner Production*, *228*, 509–520.

Mutemeri, N., Walker, J. Z., Coulson, N., and Watson, I., 2016. Capacity building for self-regulation of the Artisanal and Small-Scale Mining (ASM) sector: A policy paradigm shift aligned with development outcomes and a pro-poor approach. *The Extractive Industries and Society*, *3*(3), 653–658.

Nkuba, B., Zahinda, F., Chakirwa, P., Murhi, I., de Haan, J. S., and Bashwira, M., 2019. L'or artisanal congolais. Analyse socio-économique et de l'utilisation du mercure. Centre d'Expertise en Gestion du secteur Minier and Agence Congolaise de l'Envrionnement, Kinshasa.

O'Driscoll, D., 2017. Overview of child labour in the artisanal and small-scale mining sector in Asia and Africa. https://opendocs.ids.ac.uk/opendocs/handle/123456789/13355.

OECD, 2018. 'Gold at the crossroads: Assessment of the supply chains of gold produced in Burkina Faso, Mali and Niger'. *Author: Roberto Sollazzo.*

Pact, 2015. Unconflicted. Making conflict-free mining a reality in the DRC, Rwanda and Burundi. https://www.pactworld.org/library/unconflicted-conflict-free-mining-drc-rwanda-and-burundi.

Pathfinders for Peaceful, Just and Inclusive Societies, 2019. *The Roadmap for Peaceful, Just and Inclusive Societies – a CALL to Action to Change Our World.* New York: Center on International Cooperation. http://www.SDG 16.plus.

Rustad, S., Lujala, P., and Le Billon, P., 2012. Building or spoiling peace? Lessons from the management of high-value natural resources. In: ed. Lujala, P., Rustad, S. A. (Eds.), *High-Value Natural Resources and Peacebuilding.* London: Earthscan.

Rustad, S. A., Østby, G., and Nordås, R., 2016. Artisanal mining, conflict, and sexual violence in Eastern DRC. *The Extractive Industries and Society*, *3*(2), 475–484.

Siegel, S., and Veiga, M. M., 2009. Artisanal and small-scale mining as an extralegal economy: De Soto and the redefinition of "formalization". *Resources Policy*, *34*(1-2), 51–56.

Spiegel, S. J., 2015. Shifting formalization policies and recentralizing power: The case of Zimbabwe's artisanal gold mining sector. *Society and Natural Resources*, *28*(5), 543–558.

Stoop, N., Verpoorten, M., and van der Windt, P., 2019. More legislation, more violence? The impact of Dodd-Frank in the Democratic Republic of the Congo.

Stylo, M. S., De Haan, J. S., and Davis, K., 2020. Collecting, managing and translating data into national action plans for artisanal and small scale gold mining. *The Extractive Industries and Society*, *7*(1), 237–248.

Teschner, B. A., 2014. "Orpaillage pays for everything": How artisanal mining supported rural institutions following Mali's coup d'état. *Futures*, *62*, 140–150.

UN Environment, 2009. From conflict to peacebuilding. The role of natural resources and the environment. Geneva: UN env0069ronment. http://wedocs.unep.org/handle/20.500.11822/7867.

UN Environment, 2012. Analysis of formalization approaches in the artisanal and small-scale gold mining sector based on experiences in Ecuador, Mongolia, Peru, Tanzania and Uganda. https://wedocs.unep.org/bitstream/handle/20.500.11822/11357/Formalization_Document_Final_June_2012.pdf?sequence=1andisAllowed=y.

UN Environment, 2019. Defining the road ahead: Challenges and solutions for developing and implementing national action plan to reduce mercury use in Artisanal and Small Scale Gold Mining. Author: J. S. de Haan.

UNDP, 2016. *Mapping Mining to the Sustainable Development Goals: A Preliminary Atlas.* New York: United Nations Development Program.

United Nations, 2002. *Final Report of the Group of Experts on the Democratic Republic of the Congo.* New York: United Nations (16 October 2002. S/2002/1146).

United Nations, 2018. *Final Report of the Group of Experts on the Democratic Republic of the Congo.* New York: United Nations (4 June 2018. S/2018/531).

United Nations Economic and Social Council (ECOSOC), Progress towards the sustainable development goals: Report of the secretary-general, UN Doc E/2017/66. New York: ECOSOC. https://unstats.un.org/sdgs/files/report/2017/secretary-general-sdg-report-2017--EN.pdf.

United States Institute of Peace, 2007. *Natural Resources, Conflict, and Conflict Resolution.* Washington, DC: United States Institute of Peace. https://www.usip.org/sites/default/files/file/08sg.pdf.

Verbrugge, B., 2015. The economic logic of persistent informality: Artisanal and small-scale mining in the Southern Philippines. *Development and Change*, *46*(5), 1023–1046.

Verbrugge, B., and Besmanos, B., 2016. Formalizing artisanal and small-scale mining: Whither the workforce? *Resources Policy*, *47*, 134–141.

Verbrugge, B., and Geenen, S., 2019. The gold commodity frontier: A fresh perspective on change and diversity in the global gold mining economy. *The Extractive Industries and Society*, *6*(2), 413–423.

Westing, A. H., 1986. *Global Resources and International Conflict: Environmental Factors in Strategy Policy and Action.* Oxford: Oxford University Press.

Whaites, A., 2016. Achieving the Impossible: Can we be SDG 16 believers? *GovNet Background Paper No2, 2016*, OECD.

World Bank, 2009. *Mining Together: Large-Scale Mining Meets Artisanal Mining - A Guide for Action.* Washington, DC: World Bank.

World Bank, 2019. *State of the Artisanal and Small-Scale Mining Sector.* Washington, DC.: World Bank.

# GOAL 17 Strengthen the Means of Implementation and Revitalize the Global Partnership for Sustainable Development

*Jennifer Peyser and Stephen D'Esposito*

## CONTENTS



## SUMMARY

- Partnerships have significant potential to support all SDGs and address challenges of limited resources and resource competition among SDG planning and implementation processes.
- At a site level and globally, the mining and minerals sector has instructive experiences and examples of responding to conflict and crises with collaboration and partnerships to enhance sustainable development.

- Minerals value chains have particularly catalyzed sustainable development partnerships.
- The mining sector experience also informs opportunities and needs for partnerships to evolve and innovate to better support and scale up SDG implementation:
  - Addressing the proliferation of and need for efficiencies across voluntary sustainability initiatives
  - Measuring and managing for impact and accountability
  - Designing for more equitable engagement
  - Integrating financial sustainability and a business model for partnerships

## ANALYSIS OF MINING CONTRIBUTION TO SDG 17

### The Potential of Partnerships to Support the SDGs and Manage Resource Conflict and Competition

The sustainable development goals (SDGs) are meant to be accomplished as a comprehensive and holistic Sustainable Development Agenda for 2030. They are goals that can be complementary, mutually reinforcing, and aligned in principle.

But SDGs are not exempt from the realities of planning and decision-making processes which determine public and private financing, landscape-level planning, site permits, national laws, and international agreements. Processes to set the rules and allocate resources are typically marked by asymmetries in access and power, and a context of competition for limited resources. Unfortunately, but also unsurprisingly, this can mean that efforts to further one SDG may in fact be competing for attention and resources with other impact areas.

Siloed decision-making can also limit progress and creative thinking to scale up progress. Aid agencies, foundations, impact investors, and other public and private institutions define specific issue areas and parameters for investment; this supports a coherent portfolio with measurable impacts. However, a focus on only individual institutional goals will overlook opportunities for comprehensive approaches that link sustainability goals, such as climate and clean water, energy access and ending poverty, gender quality and economic growth. Further, SDGs cannot be achieved by a series of institutional investments alone but rather by integrating sustainable development objectives and indicators within existing business models – or developing new markets around sustainable development (Columbia Center on Sustainable Investment, UNDP, UN Sustainable Development Solutions Network, World Economic Forum, 2016).

In the face of these challenges and realities, SDG 17 encourages action to "strengthen the means of implementation and revitalize the global partnership for sustainable development." Partnership, in many of its forms, provides a platform, mechanism, and set of practices to work more holistically across sustainability goals.

In this way, partnership is an SDG, and it is also a critical mechanism to achieve the suite of SDGs and overall Sustainable Development Agenda for 2030.

While partnerships are not a perfect tool and cannot relieve the problem of limited resources and narrow scope, they may be a vehicle for managing resource competition, siloes, and transactional or zero-sum approaches to decision-making, by bringing together influencers, rights holders, and other impacted parties across sectors, geographies, and interest groups.

Partnerships can also enhance engagement in and transparency of decisions to make necessary trade-offs, by allowing parties to (1) negotiate decision-making criteria and processes for allocating limited resources; (2) apply these criteria to make decisions; and (3) define and monitor indicators to measure progress.

Further, partnerships can be additive, a layer of voluntary effort on top of legal frameworks. There is sound scepticism about the ability of voluntary initiatives alone to create the conditions necessary for sustainable development; but partnerships can fill gaps in areas of weak governance, and can provide reputational incentives to create a leadership class in the sector.

## Collaborative Responses to Site-Level and Global Conflict and Crises to Enhance Sustainable Development in the Mining and Minerals Sector

With the nonlinear pursuit of sustainability, partnerships are needed both for progress and in the face of regression, and the mining sector provides historical and current examples of both to inform those working across other SDGs (RESOLVE and World Economic Forum, 2015; Kickler and Franken, 2017; IISD, 2018).

At any mine site there is an array of perspectives on how to prioritize and pursue community engagement and consent; environmental, cultural, and social outcomes; economic benefits; land and mineral rights; labor and working conditions; and other critical issues—giving rise to conflicts in all possible combinations between and among communities, companies, and governments. These conflicts also scale up to national, regional, and global levels as linkages are made between communities, mining operations, laws and conventions, and other fora for global governance.

In response to conflicts stemming from mining operations, the industrial mining and minerals sector has over time undertaken research, dialogue, and independent evaluation to define and consider good practices to improve environmental, social, and governance outcomes at the mine-site level and in impacted communities. These efforts to understand, define, implement, and measure sustainability impacts—on a proactive basis as well as in response to failures—are an important foundation for the sector's contributions to partnerships for sustainable development.

Two important examples—including a proactive effort at the global level and a crisis response at a site level—are instructive of how expertise has evolved and created a foundation for partnership in sustainable development:

**Global action on cross-cutting sustainability outcomes: MMSD and the ICMM.** The Mining, Minerals and Sustainable Development Project (MMSD) was an early, successful partnership effort. MMSD supported research on how the mining and minerals sector could contribute to the global transition to sustainable development. Fifteen years later, MMSD remains an important substantive and process reference point. While stakeholders in our networks hold differing views on whether or not MMSD was a success most agree that it successfully reset expectations for the type of partnerships essential to industry progress. The International Council on Mining and Metals (ICMM) was founded as an outcome of the MMSD project, and the platform to take this work forward. We can point to the positive substantive outcomes of ICMM such as position statements on water management: security; free, prior and informed consent (FPIC); and recent leadership in the independent tailings review, all of which help support industry partnerships on SDGs.

**Site-level response to crisis and conflict: Community Relationships Review.** In 2007, Newmont Mining Corporation initiated a Community Relationships Review (CRR) in response to site-based issues and concerns raised by stakeholders and shareholders (Smith and Feldman, 2009). Governance and guidance of the CRR included an independent stakeholder advisory panel, and independent researchers were commissioned to review community relationships at a number of Newmont's sites (Kemp, Evans, Plavina, and Sharp, 2008). The CRR was a unique, creative company response in a highly charged environment. It created a sense of shared responsibility across stakeholder groups and the company for an outcome that led to improvement.

With a variety of types and intentions of actors in the mining and minerals sector, and the ongoing challenges of applying global goals at a site level, conflict continues in the mining sector and social license to operate is understood as a process rather than an endpoint. In fact, conflicts are sometimes driven by major impacts to communities and ecosystems, such as tailings dam failures, public and private security incidents, and corruption. As in other sectors, there remains a role and need for adversarial processes as a means to remedy some impacts and injustices. The informal, artisanal, and small-scale mining sector has additional complexity, sometimes with a shared footprint with large-scale mining but often with a unique set of partnership interventions required.

Whether in response to conflict or through a proactive, collaborative framework, the mining sector demonstrates the value of mobilizing stakeholders and rights holders in resolving crises and defining a path forward. To support this, an increasingly sophisticated community of practice, including community members, non-governmental organizations, and technical experts has developed over time to respond to critical issues and incidents. The 2019–2020 Global Tailings Review, convened by ICMM, the United Nations Environment Programme, and the Principles for Responsible Investment, provides a current example of a multi-stakeholder response to the major environmental disaster of the collapse of the Fundão tailings dam and other failures that preceded it. The initiative has the aim of creating an international standard which "creates a step change for the industry in the safety and security of tailings storage facilities" with a process engaging an independent chair and diverse stakeholder engagement, to complement technical industry experts (UNEP, PRI, and ICMM, 2019). This process has the potential to be another example of the value of partnership in both pursuing sustainability goals and correcting course in the face of crisis.

### Mineral Value Chains as Catalyst and Platform for Expanded Partnership and Impact

Shared supply chain accountability for natural resources, including mining, is a growing area of partnership in the sector. The emergence of risk assessment and due diligence in global minerals supply chains has built on and deepened the need and value for partnerships in addressing sustainable development in the mining and minerals sector. Heightened scrutiny of consumer brands, including demand for information about suppliers and sourcing practices by multinational companies, have come from government, civil society, and business and other industry partners.

Stakeholders from all sectors are seeking more information and better performance on economic, environmental, labor, health, governance, and other factors. This demand for due diligence, and the systems that have emerged to enable due diligence, have forged or made more transparent both material and reputational linkages—and risks—between producers and consumers in developed nations and the economies, resources, and people of mineral-producing countries, sometimes down to the mine-site level.

In this way, responsible mining and responsible sourcing has meant an expansion of communications, information-sharing, and problem-solving from the mine through to industrial and consumer products, thus increasing the reach of the minerals and mining sector and its opportunity to contribute to sustainable development through partnerships. These mining-related partnerships, and resources generated to support these partnerships, touch many facets of the SDG 17 targets, including finance, technology, capacity building, trade; and these partnerships particularly serve as a vehicle to address systemic issues that require tripartite, local to global, and cross-sector engagement.

Responsible sourcing activities have also led to a suite of voluntary efforts to complement or even precede legal and regulatory frameworks, with many focused on the ASM sector. For example, the *OECD Due Diligence Guidance for Responsible Supply Chains of Minerals from Conflict-Affected and High-Risk Areas* was adopted through a multi-stakeholder consultation process, and now serves as an alignment tool for a suite of industry initiatives, regulatory efforts, and regional governance initiatives. (For a selected set of examples of active partnership and multi-stakeholder models and example initiatives, please see Table 17.1.)

## EVOLUTION AND INNOVATIONS IN MINING AND MINERALS PARTNERSHIPS TO INCREASE THE SCALE AND PACE OF CHANGE

In response to the shared interest of communities, government, investors, and supply-chain actors to reduce the negative impacts of natural resource extraction and improve community and broader social benefits, a wide range of partnerships have been designed and implemented over the past two to three decades. The mining sector and its stakeholders can now reflect on progress, gaps, and opportunities for enhanced coordination to scale up sustainability progress, prevent site-level setbacks, and consider sector-wide progress on the SDGs and ways in which partnerships can increase the scale and pace of implementation.

**TABLE 17.1**
**Selected Examples of Currently Operating (2019–2020) Partnership Models and Specific Initiatives on Responsible Mining and Responsible Minerals Sourcing That Contribute to SDG 17 Targets**

| SDG 17 Target | Selected SDG 17 Indicators | Example Mining Sector Partnership Models | Selected Example Initiatives |
|---|---|---|---|
| **Finance** | Strengthen domestic resource mobilization, including through international support to developing countries, to improve domestic capacity for tax and other revenue collection | **Partnerships for revenue reporting and improved transparency practices** | • Extractives Industry Transparency Initiative |
| | Mobilize additional financial resources for developing countries from multiple sources | **Public–private partnerships for financial resources in developing countries** | • European Partnership for Responsible Minerals<br>• Public–Private Alliance for Responsible Minerals Trade |
| | | **Multi-donor funds for responsible mining projects in developing countries** | • planetGOLD |
| **Technology** | Promote the development, transfer, dissemination, and diffusion of environmentally sound technologies to developing countries on favorable terms | **Partnerships for mercury reduction** | • Global ASM Challenge<br>• Global Mercury Partnership<br>• planetGOLD |
| | | **Partnerships for complementary measures to support economic development and stability in mineral-producing, developing countries** | • Congo Power<br>• International Cyanide Management Code<br>• Intergovernmental Forum on Mining, Minerals, Metals (IGF) |

*(Continued)*

**TABLE 17.1 (CONTINUED)**
**Selected Examples of Currently Operating (2019–2020) Partnership Models and Specific Initiatives on Responsible Mining and Responsible Minerals Sourcing That Contribute to SDG 17 Targets**

| SDG 17 Target | Selected SDG 17 Indicators | Example Mining Sector Partnership Models | Selected Example Initiatives |
|---|---|---|---|
| **Capacity Building** | Enhance international support for implementing effective and targeted capacity-building in developing countries to support national plans to implement all the sustainable development goals, including through North–South, South–South and triangular cooperation | **Responsible mining or due diligence standards and certification systems, and the suite of capacity building for artisanal/small-scale and industrial mining operations that accompanies these systems** | • CRAFT<br>• Fairmined<br>• Fair Trade<br>• ICMM Performance Expectations<br>• Initiative for Responsible Mining Assurance<br>• iTSCi<br>• Toward Sustainable Mining |
| | | **Government-to-government and regional partnerships** | • "Support for the implementation of the 2030 Agenda for Sustainable Development in Latin America and the Caribbean" (ECLAC-Germany)<br>• ICGLR Regional Initiative against the Illegal Exploitation of Natural Resources<br>• Global Battery Alliance |

*(Continued)*

**TABLE 17.1 (CONTINUED)**
**Selected Examples of Currently Operating (2019–2020) Partnership Models and Specific Initiatives on Responsible Mining and Responsible Minerals Sourcing That Contribute to SDG 17 Targets**

| SDG 17 Target | Selected SDG 17 Indicators | Example Mining Sector Partnership Models | Selected Example Initiatives |
|---|---|---|---|
| **Trade** | Significantly increase the exports of developing countries, in particular with a view to doubling the least developed countries' share of global exports by 2020 | **Global value-chain due-diligence systems and related multi-stakeholder consultations and partnerships** | • Responsible Jewellery Council Code of Practices and Chain of Custody Certification<br>• LBMA Responsible Gold Guidance<br>• OECD Due Diligence Guidance for Responsible Sourcing of Minerals from Conflict-Affected and High-Risk Areas<br>• Responsible Minerals Assurance Program |
| | | **Individual/targeted supply chain or "closed-pipe' efforts: Responsible sourcing supply-chain projects aimed at increasing legal/formal trade from least developed countries** | • Better Gold Initiative<br>• Just Gold<br>• Kimberley Process<br>• Responsible Artisanal Gold Solutions Forum<br>• USAID Conflict Free and Commercially Viable Gold Program<br>• planetGOLD country projects |
| | | **Investment-focused initiatives** | • Solidaridad Accelerator for Responsible Gold<br>• Impact Facility for Sustainable Mining Communities<br>• planetGold Access to Finance<br>• CRAFT Impact Marketplace |

*(Continued)*

**TABLE 17.1 (CONTINUED)**
**Selected Examples of Currently Operating (2019–2020) Partnership Models and Specific Initiatives on Responsible Mining and Responsible Minerals Sourcing That Contribute to SDG 17 Targets**

| SDG 17 Target | Selected SDG 17 Indicators | Example Mining Sector Partnership Models | Selected Example Initiatives |
|---|---|---|---|
| Systemic Issues | Enhance global macroeconomic stability, including through policy coordination and policy coherence | **Multilateral and global government-to-government fora** | • OECD Fora on Responsible Mineral Supply Chains and Due Diligence Guidance;<br>• Intergovernmental Forum on Mining, Minerals, Metals and Sustainable Development (IGF) |
| | Enhance policy coherence for sustainable development | **Global fora and model guidance and supply chains as tools for alignment and coherence across systems** | • IGF<br>• OECD Due Diligence Guidance |
| | | **Issue-specific fora supporting global policy alignment and public–private implementation** | • FPIC Solutions Dialogue<br>• Voluntary Principles for Security and Human Rights |
| | | **Standards and certification systems for artisanal or industrial mining, and attendant alignment efforts** | • Aluminium Stewardship Initiative<br>• CRAFT<br>• Fairmined<br>• Fairtrade Gold<br>• ICMM Performance Expectations<br>• Initiative for Responsible Mining Assurance (IRMAs or IRMA)<br>• LBMA Responsible Gold Guidance<br>• Maendeleo Diamond Standard<br>• Towards Sustainable Mining (TSM)<br>• Responsible Jewellery Council<br>• Responsible Minerals Assurance Program<br>• Responsible Steel |

(*Continued*)

**TABLE 17.1 (CONTINUED)**
**Selected Examples of Currently Operating (2019–2020) Partnership Models and Specific Initiatives on Responsible Mining and Responsible Minerals Sourcing That Contribute to SDG 17 Targets**

| SDG 17 Target | Selected SDG 17 Indicators | Example Mining Sector Partnership Models | Selected Example Initiatives |
|---|---|---|---|
| | | **Multi-stakeholder and expert-consulted research** | – |
| | By 2020, enhance capacity-building support to developing countries, to increase significantly the availability of high-quality, timely, and reliable data disaggregated by income, gender, age, race, ethnicity, migratory status, disability, geographic location, and other relevant characteristics | **Public domain, open-source, and pooled databases, including on geologic/mineral resources, sector statistics, and geospatial data on minerals/natural resources** | • African Minerals Geoscience Initiative<br>• Delve<br>• MapX |
| | Enhance the global partnership for sustainable development, complemented by multi-stakeholder partnerships that mobilize and share knowledge, expertise, technology, and financial resources, to support the achievement of the sustainable development goals in all countries, in particular developing countries | **Inclusive of above models** | **Inclusive of above selected examples** |

*Note:* This is not meant as a comprehensive list but a sample of active initiatives. For comprehensive reviews, please see, e.g., RESOLVE and World Economic Forum, 2015; Kickler and Franken, 2017; IISD, 2018.

## Addressing Proliferation and Increasing Efficiency of Partnerships

Positive results from partnerships have spurred the development of dozens of dialogues, issue- or commodity-specific, company-focused, and site-based voluntary initiatives on mining and minerals—and observations about diminishing returns and need for enhanced coordination (Columbia Center on Sustainable Investment, UNDP, UN Sustainable Development Solutions Network, World Economic Forum, 2016). In one survey of experts and stakeholders, 96% of respondents agreed that there is potential to create linkages or efficiencies among these initiatives (RESOLVE and World Economic Forum, 2015). Recognizing both redundancies and that a number of initiatives addressed unique issues or aspects of the sector, subsequent work has focused on opportunities for enhancing alignment and interoperability among initiatives and especially multi-stakeholder standards (ISEAL Alliance, 2018).

Efforts include those by TSM, IRMA, Responsible Steel, Responsible Jewellery Council, and ISEAL to explore, test, and promote alignment among standards, including setting specific benchmarks and considering joint audits (IISD, 2018) (ISEAL Alliance, 2018) (Initiative for Responsible Mining Assurance, 2018). Other sustainability tools are also being created to promote alignment on a priority issue and support coherence in setting sustainability goals and measuring progress across the full range of industry and multi-stakeholder standards systems, such as RESOLVE's Climate Smart Emissions Reporting Widget (RESOLVE, 2019).

## Measuring and Managing for Impact and Accountability

Partnerships in the mining and minerals sector have over time created a significant asset for the industry, governments, and stakeholders. This tool chest may be currently underutilized because key actors have limited awareness or access. An inventory of SDGs tools (e.g., codes, standards, financial and technical resources, and infrastructure) could be useful to stakeholders at all levels. While partnerships will typically communicate on outcomes or progress toward objectives via annual reports, case studies, newsletters, or other vehicles, an annual survey could also prove useful.

At the same time, partnerships have widely varied expertise in and approaches to monitoring and evaluation of their work, including designing and managing for their intended sustainable development impacts. Like the SDGs, partnerships require indicators and measurable goals to accompany the high-level objectives that bring partners together. To maximize impact, partnerships should engage technical support in creating results frameworks, including defining outcomes and indicators to measure and report on progress, identify gaps, inform course correction where needed, and to hold partners accountable to their shared goals.

## Designing for More Equitable Engagement

As in any multi-stakeholder effort, the efficacy of a minerals sector partnership can be limited by its membership. Greater diversity and equity related to Indigenous peoples, gender, geography (global North–South), culture, economic status, and other elements are critical to achieving SDGs, and must be reflected in the makeup of any partnership that hopes to contribute to the SDGs.

The minerals sector has recognized the distinction between stakeholders and rights-holders, and both the importance and obligation to include Indigenous peoples and other impacted parties in consultation and partnerships at the site and global levels. Principles such as FPIC have been referenced in major industry benchmarks including the IFC Performance Standards, ICMM Performance Expectations, TSM, the Initiative for Responsible Mining Assurance (IRMA), and the Aluminium Stewardship Initiative (ASI). Groups such as First Nations Women Advocating Responsible Mining have advised on standards, such as IRMA (Initiative for Responsible Mining Assurance, 2018); and an Indigenous People's Advisory Forum and two Indigenous representatives on the Standard Committee are integrated into ASI's governance (Aluminium Stewardship Initiative, 2018). The FPIC Solutions Dialogue is another global partnership that goes beyond recognizing Indigenous rights to directly engaging rights holders in dialogue and guidance on best practice (RESOLVE, 2020).

Those convening minerals sector partnerships should further challenge themselves and strive for more meaningful engagement of Indigenous peoples, as well as inclusion of women, people of color, economically disadvantaged, politically vulnerable, and other historically disenfranchised stakeholders and rights holders as an imperative as well as a means to improving decisions and outcomes.

## INTEGRATING FINANCIAL SUSTAINABILITY AND A BUSINESS MODEL FOR PARTNERSHIPS

If partnerships are to be successful, they must be financially viable. Many mining sector partnerships have been reliant on donor funding or philanthropy for success—including a significant list of pilot projects or assurance systems that have never scaled.

While past initiatives have not necessarily been designed for long-term operation or revenue generation, a next generation of partnerships for sustainability must consider self-financing mechanisms or other innovations that acknowledge the need for long-term solutions and therefore a long-term business plan. Examples including establishment of social enterprises, impact investments, service provision, and other models should be further designed and tested in the mining sector, for industrial and artisanal or small-scale mining, responsible sourcing, and other minerals-related sustainability objectives.

## CASE ANALYSIS

**Susan Keane**

### PLANETGOLD: A GLOBAL PARTNERSHIP FOR REDUCING MERCURY USE IN ASGM

Nearly two decades ago, the international scientific community identified mercury as a dangerous global pollutant and called for coordinated action among nations to reduce the significant health and environmental risks posed by this powerful neurotoxin. Subsequent assessments found that the largest source of mercury pollution in the world is artisanal and small-scale gold mining (ASGM). Drawing on existing knowledge, experience, and robust debate in the development community on artisanal and small-scale mining (ASM), a voluntary coalition of governments, NGOs, and others began to promote projects to reduce mercury use in the sector, including through the Global Mercury Partnership convened under the United Nations Environment Programme (UNEP). In 2009, member countries of UNEP agreed to negotiate a legally binding instrument that became the Minamata Convention on Mercury. When crafting the text of this treaty, negotiators recognized the unique nature of ASGM as both an environmental and economic development issue and included mandatory provisions that require parties to address ASGM holistically, through technical, social, educational, and financial strategies.

Since the Minamata Convention on Mercury entered into force in 2017, funding has begun to flow to support parties to meet their obligations. One of the first initiatives approved from the treaty's financial support mechanism was the planetGOLD programme. Funded by the Global Environment Facility, the programme aims to reduce mercury use in ASGM in eight countries. Following the obligations of the Minamata Convention for addressing mercury use in ASGM, projects in each planetGOLD country share a common, comprehensive design, addressing technology transfer, access to finance and markets, and formalization of the sector.

The planetGOLD programme directly responds to Sustainable Development Goal 17, which aims to "strengthen the means of implementation and revitalize the global partnership for sustainable development." In particular, per Subgoal 17.7, planetGOLD promotes development and dissemination of mercury-free technologies and is experimenting with mechanisms to improve access to finance so that miners can afford these technologies, while recognizing the realities of miners on the ground. In addition to working at the country level, the planetGOLD programme also includes a Global Project to mobilize and share knowledge among the countries participating in the programme, as well as across the global community of practice on ASGM, as envisioned in Subgoal 17.16. Finally, the programme is serving as a platform to connect government, miners, financial institutions, technical specialists, gold buyers, and NGOs to come together create practical solutions for the sector, as envisioned in Subgoal 17.17.

## REFERENCES

Aluminium Stewardship Initiative. (2018). Brochure on the Indigenous People's Advisory Forum. Retrieved from Aluminium Stewardship Initiative: https://aluminium-stewardship.org/about-asi/indigenous-peoples/.

Columbia Center on Sustainable Investment, UNDP, UN Sustainable Development Solutions Network, World Economic Forum. (2016, July). Mapping Mining to the Sustainable Development Goals: An Atlas. Retrieved from https://www.undp.org/content/undp/en/home/librarypage/poverty-reduction/mapping-mining-to-the-sdgs--an-atlas.html.

IISD. (2018). State of Sustainability Initiatives Review: Standards and the Extractive Economy. Retrieved from IISD: https://www.iisd.org/sites/default/files/publications/igf-ssi-review-extractive-economy.pdf.

Initiative for Responsible Mining Assurance. (2018). First Nations Women Advocating for Responsible Mining. Retrieved from Responsible Mining Map: https://map.responsiblemining.net/company/75.

Initiative for Responsible Mining Assurance. (2018). Towards Sustainable Mining. Retrieved from Responsible Mining Map: https://map.responsiblemining.net/company/87.

ISEAL Alliance. (2018, July). Creating Shared Value: Summary Report. Retrieved from ISEAL: https://www.isealalliance.org/sites/default/files/resource/2018-07/ISEAL_Creating%20Shared%20Value_summary%20report_July%202018.pdf.

Kemp, D., Evans, R., Plavina, J., and Sharp, B. (2008, May). Newmont's Global Community Relations Review: Organisational Learnings from the Minahasa Case Study. Retrieved from Newmont Goldcorp: https://www.newmontgoldcorp.com/wp-content/uploads/2019/05/CRR-Minahasa-Report.pdf.

Kickler, K., and Franken, G. (2017, March). Sustainability Schemes for Mineral Resources: A Comparative Overview. Retrieved from Bundesanstalt für Geowissenschaften und Rohstoffe (Federal Institute for Geosciences and Natural Resources): https://www.bgr.bund.de/EN/Themen/Min_rohstoffe/Downloads/Sustainability_Schemes_for_Mineral_Resources.pdf?__blob=publicationFile&v=6.

RESOLVE. (2019, December). Climate Smart Mining Emissions Widget. Retrieved from RESOLVE: https://www.resolve.ngo/climate_smart_widget.htm.

RESOLVE. (2020). FPIC Solutions Dialogue. Retrieved from RESOLVE: https://www.resolve.ngo/fpic.htm.

RESOLVE and World Economic Forum. (2015, August). Voluntary Responsible Mining Initiatives: A Review. Retrieved from World Economic Forum: http://www3.weforum.org/docs/Voluntary_Responsible_Mining_Initiatives_2016.pdf.

Smith, G. A., and Feldman, D. (2009, March). Community Relationships Review: Global Summary Report. Retrieved from Newmont Goldcorp: https://www.newmontgoldcorp.com/wp-content/uploads/2019/05/CRR-Global-Summary-Report-and-Appendices-English.pdf.

UNEP, PRI, and ICMM. (2019, June). Independent Review of Global Best Practices to Inform an International Standard on Mine Tailings Storage Facilities. Retrieved from Global Tailings Review: https://globaltailingsreview.org/about/.

# Index

## R



## S



## T



## U



## W



## Z

Taylor & Francis Group
an **informa** business

# Taylor & Francis eBooks

www.taylorfrancis.com

A single destination for eBooks from Taylor & Francis with increased functionality and an improved user experience to meet the needs of our customers.

90,000+ eBooks of award-winning academic content in Humanities, Social Science, Science, Technology, Engineering, and Medical written by a global network of editors and authors.

## TAYLOR & FRANCIS EBOOKS OFFERS:

- A streamlined experience for our library customers
- A single point of discovery for all of our eBook content
- Improved search and discovery of content at both book and chapter level

## REQUEST A FREE TRIAL

**support@taylorfrancis.com**

Routledge
Taylor & Francis Group

CRC Press
Taylor & Francis Group

For Product Safety Concerns and Information please contact our EU representative GPSR@taylorandfrancis.com
Taylor & Francis Verlag GmbH, Kaufingerstraße 24, 80331 München, Germany

www.ingramcontent.com/pod-product-compliance
Ingram Content Group UK Ltd.
Pitfield, Milton Keynes, MK11 3LW, UK
UKHW011006180926
475701UK00004B/593

*9780367553838*